Photoacoustic Imaging - Principles, Advances and Applications

Edited by Reda R. Gharieb

Published in London, United Kingdom

IntechOpen

Supporting open minds since 2005

Photoacoustic Imaging - Principles, Advances and Applications
http://dx.doi.org/10.5772/intechopen.75344
Edited by Reda R. Gharieb

Contributors
Surya Narayan Thakur, Yannis Paulus, Yanxiu Li, Elena Petrova, Sharmila Anandasabapathy, Sheena R. Bhushan, Chin Hsia, Igor Romanyshyn, Rostyslav Romanyshyn, Galyna Romanyshyn, Reda Gharieb

First published in London, United Kingdom, 2020 by IntechOpen
IntechOpen is the global imprint of INTECHOPEN LIMITED, registered in England and Wales, registration number: 11086078, 7th floor, 10 Lower Thames Street, London, EC3R 6AF, United Kingdom
Printed in Croatia

British Library Cataloguing-in-Publication Data
A catalogue record for this book is available from the British Library

Additional hard and PDF copies can be obtained from orders@intechopen.com

Photoacoustic Imaging - Principles, Advances and Applications
Edited by Reda R. Gharieb
p. cm.
Print ISBN 978-1-78984-766-6
Online ISBN 978-1-78984-767-3
eBook (PDF) ISBN 978-1-78984-623-2

For EU product safety concerns:
IN TECH d.o.o., Prolaz Marije Krucifikse Kozulić 3, 51000 Rijeka, Croatia, info@intechopen.com or visit our website at intechopen.com.

Meet the editor

Reda R. Gharieb PhD, is Professor of Biomedical and Bioelectronics Engineering at Assiut University, Egypt. He has worked for Fairway Medical Technologies Inc. and Seno Medical Instruments Inc., Texas, USA, on their photoacoustic imaging (PAI) technology. He has developed algorithms for 2D image reconstruction in PAI of breast and prostate cancers. He also has developed an algorithm for 3D image reconstruction in PAI of a small animal, using a rotated ARC-shaped sensor array. Dr. Gharieb has collaborated with brilliant international professors, scholars and engineers from academic and industrial institutions. He authored two books, three chapters, two patents and about 60 papers. His research interests include signal/image processing, modeling and simulation, machine learning, statistical and scientific computing, neuro engineering, bioelectronics and bioinstrumentation.

Contents

Preface

Photoacoustic imaging (PAI) is an emerging non-invasive imaging modality that integrates the advantages of deep ultrasound penetration and high optical contrast. The fundament of PAI is photoacoustic effect, which takes place when a light-absorbing material sample is irradiated by a laser pulse, causing temporal changes of the temperature at loci where light energy is absorbed. Due to this temperature change, the region where light energy is absorbed exhibits transient thermoplastic expansion, generating a pressure pulse. This pressure pulse propagates within the sample body to the surrounding areas and is detected by a set of ultrasonic sensors. The amplitude, time of arrival at the sensor, and duration of each detected pulse are proportional to the absorption coefficient, the location, and the size of the vibrating region, respectively. The set of detected ultrasonic signals are used by different image reconstruction algorithms to produce an image that maps the spatial distribution of the light absorption coefficient within the material sample. Because it is nonionizing, nondestructive, and high-resolution, PAI is opening new frontiers in medical imaging. It provides diagnostic imaging for the anatomy and physiology of small animals, anatomical tissue/organ structure, oxygenation status, blood flow, abnormal morphology, functional vasculature characteristics, skin burns, hypermetabolism, and vulnerable plaques in the coronary arteries.

This book, *Photoacoustic Imaging: Principles, Advances and Applications*, provides interested readers with the principle knowledges, advanced methodologies, and new applications associated with PAI technology. The first section, which contains four chapters, covers the generation and detection of photoacoustic signals, sound source localization, and image reconstruction and formation. This section cites different research papers on emerging deep learning-based image reconstruction. In the second section, which contains two chapters, presents relatively new diagnostic PAI applications. The first chapter discusses PAI application in gastroenterology. It explores the clinical use of PAI-based endoscopy in the diagnosis of esophageal cancer, inflammatory bowel diseases, and pancreatic cancer. The final chapter reports that many eye diseases, including macular degeneration and diabetic retinopathy, involve abnormalities in the vasculature. Therefore, PAI for visualizing and investigating the vasculature can be incredibly helpful as a diagnostic tool for eye diseases. These include iris vasculature, corneal neovascularization, retinal vasculature, choroidal vasculature, and retinal neovascularization. PAI is also used to visualize retinal metabolism and melanin of the eye.

We believe that the applications of PAI in gastroenterology and eye diseases are relatively new and deserve attention. In the future, there is sure to be more literature on using PAI in other areas. We hope this book serves as a channel for exchanging perspectives, experiences, and research results within the scientific community.

Finally, the editor acknowledges and appreciates the effort and kind cooperation of all the chapter authors and their willingness to exchange their experiences and research outcomes with other colleagues. The editor also acknowledges and appreciates the important editing and reviewing services, strong support, and kind cooperation from the staff at IntechOpen: Ms. Klara Mestrovic, Commissioning Editor; Ms. Nina Kalinic Babic, Author Service Manager; and Mr. Ivan Butkovic, Author Service Manager.

Reda R. Gharieb
Assiut University,
Egypt

Section 1

Photoacoustic Imaging: Principles and Advances

Chapter 1

Photoacoustic Imaging for Cancer Diagnosis: A Breast Tumor Example

Reda R. Gharieb

Abstract

Photoacoustic (PA) imaging utilizes laser pulses to deliver energy to an examined object for the generation of ultrasonic waves. Thus, it provides a noninvasive and nonionizing imaging modality. Therefore, it has found clinical use for cancer diagnosis in different organs, e.g., breast, prostate, and thyroid nodules. It offers morphological, functional, and molecular imaging. Moreover, the oxygen saturation in a body can be computed by calculating the wavelength-dependent light absorption coefficients at two different wavelengths. In this chapter, the principle of the PA imaging is introduced for the present book.

Keywords: photoacoustic imaging, backprojection, breast tumor, ARC-shaped ultrasonic detectors

1. Introduction

Photoacoustic (PA) imaging is an emerging diagnostic modality that gets advantage of the optically induced ultrasonic signals in tissues [1–5]. PA imaging in cancer diagnosis relies on the enhanced optical absorption of tumors and the relatively high optical transparency of normal tissues in addition to low acoustic distortion and attenuation of tissues [6]. Cancer cells gradually develop a dense microvascular network, which appears to be a marker that a tumor is aggressively growing and subject to metastasis. Furthermore, not only the amount of blood is substantially higher in malignant tumors compared to normal tissue, but also the blood in malignant tumors tends to be less oxygenated blood than the one in normal tissue. Those abnormalities of blood amount and plus being less oxygenated increase the optical absorption of the cancer tissue in the near-infrared (NIR) region.

PA imaging has been combined with ultrasound imaging and augmented with molecular targeted contrast agents [7, 8]. This makes this hybrid modality capable of imaging cancer at the cellular and molecular level, thus opening diverse opportunities to improve diagnosis of tumors, detect circulating tumor cells, and identify metastatic lymph nodes.

PA imaging consists of shining the object of interest, e.g., prostate/breast by a laser pulse; the light energy absorbed by the object tissues causes a thermal induced mechanical vibration within the tissues. The intensity of the mechanical vibration is proportional to the light absorption coefficient of different region within the tissues. The mechanical vibration generates ultrasonic waves that propagate outward the object [1, 6, 9]. Theses ultrasonic signals can be recorded at different locations on the object surface and are used to construct an image for the absorption coefficient of the tissue region within the object. This distribution of the light absorption coefficient within the object demonstrates the tumor location and shape. The image construction, which provides an inverse problem solution, can be done analytically or by computed tomography through different backprojection methods [10].

In the next subsections, a review of a PA imaging system is presented. The PA imaging for the breast cancer detection is used as an example. The review highlights the array of the ultrasonic detectors, the received signals, and the formation of the image using a weight and sum backprojection algorithm.

2. General PA imaging system

Figure 1 shows a block diagram of the photoacoustic imaging principle. As shown, laser pulse is used to deliver optical energy to an absorptive object, e.g., breast/prostate tissues, whole body of a small animal. The tissues absorb some energy from the laser light and get heated up. After heating, the tissues expand or vibrate in order to release the absorbed heat energy. This sudden and fast expansion of tissues in turn causes sound waves to be generated and they are in the frequency range of Mega Hertz (Ultrasound). These ultrasonic waves are recorded using a set of piezoelectric detectors. The time taken to receive ultrasound after the laser pulse has been given is used to find the location of the tissues which vibrate. Thus, the ultrasonic signals can be handled so that to provide a morphological image in two- or three-dimensional for the absorptive object. The image formation is accomplished utilizing different backprojection methods [10]. These methods take into account the light distribution, the geometry of the ultrasonic detectors array, and the directivity pattern of each detector. In [11], the recent advancement of light-emitting diode (LED) technology has shown to provide a less expensive and more safety source of light to replace the class-IV laser systems. Recently, integrating a microrobotic system and PA imaging has enabled deep imaging and precise control of the micromotors in vivo [12].

Near-infrared (NIR) light (wavelength range of 700–2500 nanometers) is used as a laser pulse to deliver energy to the tissues. This is because of the fact that NIR can penetrate skin since the skin absorbs very less light at longer wavelengths. However, as NIR skips past the skin to deeper tissues, water and other components in deeper tissues absorb NIR and therefore get heated up and release the ultrasonic waves.

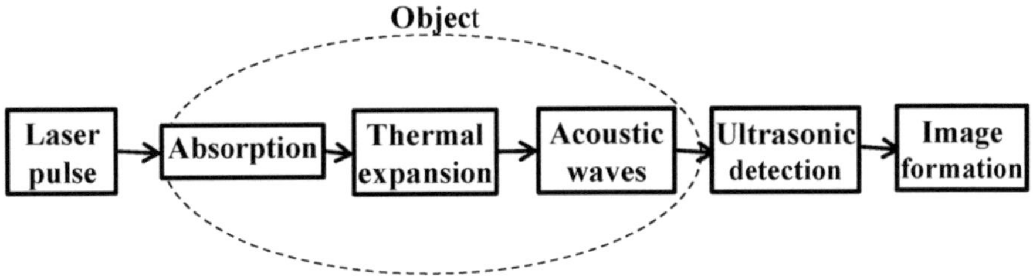

Figure 1.
A general block diagram of PA imaging principle.

3. Advantages and disadvantages of PA imaging

Advantages include:

1. nonionizing imaging since it relies on just NIR light (optical radiation) and ultrasound;

2. high contrast and good spatial resolution. The high contrast is due to the use of the optical contrast absorption and the good spatial resolution is due to the use of ultrasonic signals resolution. Thus, it is a single hybrid imaging modality;

3. provides better depth of penetration than different optical imaging and better spatial resolution than ultrasonic imaging; and

4. nanoparticles can be used to enhance the tissue heating process and therefore the ultrasonic signals and the final image.

Disadvantages include:

1. depth of penetration and distribution of light are issues.

4. Ultrasound signals and backprojection

As illustrated by **Figure 2**, the ultrasonic pressure wave received at a circle of radius r_0 in the 2D space due to a laser pulse is given by [13]

$$p(r_0, t, \varphi) = \frac{v_s B I_0}{4\pi C} \frac{\partial}{\partial t} \oint_{|r_0 - r| = v_s t} \frac{A(r)}{|r_0 - r|} dr \tag{1}$$

where v_s is the speed of the acoustic waves; B is the coefficient of volumetric thermal expansion; C is the specific heat capacity; I_0 is a scaling factor proportional to the incident radiation intensity; and $A(r)$ describes the to-be-reconstructed electromagnetic absorption property of the medium at r, given $p(r_0, t)$.

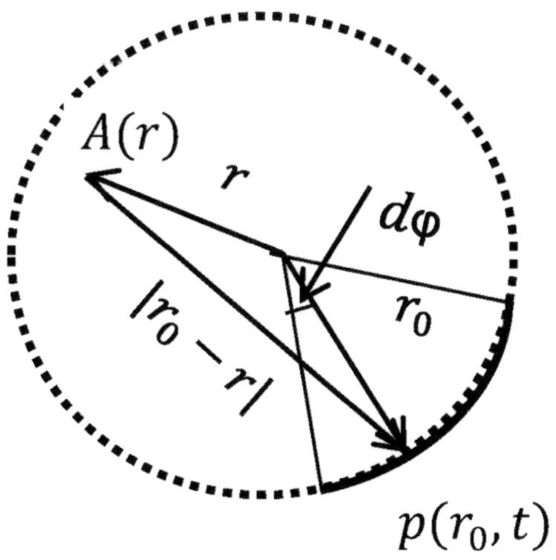

Figure 2.
The acoustic pressure wave $p(r_o, t)$ detected on an ARC of radius r_o and angle φ due to the electromagnetic absorption $A(r)$.

For 2D image formation using the recorded acoustic pressure $p(r_0, t)$, the approximate inverse solution is given by [14]

$$A(r) = -\frac{r_0^2 C}{4\pi B I_0 v_s^4} \int_{\varphi} \frac{1}{t} \frac{\partial p(r_0, t)}{\partial t} \bigg|_{t=|r_0 - r|/v_s} d\varphi \tag{2}$$

This implies that first, compute time-derivative of the pressure wave, divide the resultant by t, and finally, integrate over the ARC angle. Notice that r is determined by the time the wave taken to travel from the point r to r_0.

Thus, practically to detect the ultrasonic signals released form the absorptive object excited by the laser pulse, a set of the piezoelectric sensors are used. These sensors are arranged in an array of a certain geometry either linear, surface, curvature, etc. The number of sensors, the geometry of the array, and the directivity of each sensor, the ability of the sensor to receive a signal from only a certain direction, are important factors for the image quality. Each sensor is followed by a low-noise amplifier to amplify the detected signal. **Figure 3** shows the geometry of an ARC-shaped array of 64 sensors that has been adopted for breast cancer imaging [6]. These sensors are uniformly distributed on an angle of 166°. All the sensors have the directivity, the maximum gain, to the center of the ARC, the point $(0, 0)$. Thus, the recorded signal by the kth sensor in time-domain can be expressed by

$$v_k(t), t = 0, T, 2T, 3T, \ldots \tag{3}$$

where T is the sampling time.

Assuming uniform light distribution within the tissue, the back projection of the kth sensor signal is given by

$$G_k(x, y) = \overline{v_k}\left(\frac{R\left(x - x_{k,}, y - y_{k,}\right)}{v_s}\right) R\left(x - x_{k,}, y - y_{k,}\right) \cos\left(\theta_k(x, y)\right), \ -7 < x < 7, \ -4 < y < 0 \tag{4}$$

where $\overline{v_k}$ is the velocity potential computed by the time-integration of the received pressure signals; and $v_s = 1540$ m/ sec is the speed of sound in tissues.

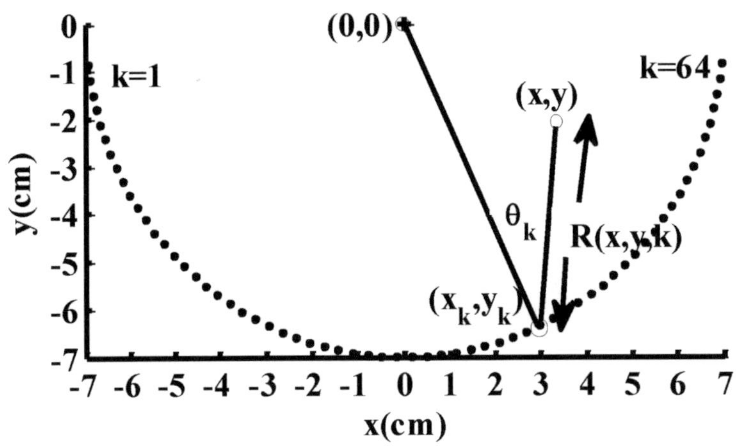

Figure 3.
ARC-shaped PA ultrasonic detecting array.

We can also restrict the projected location (x,y) for only $\{(x,y) : |\theta_k(x,y)| \leq \frac{\pi}{2}\}$. For the M sensors, the back projection is done by the summation of (4) over the M sensors, that is

$$G(x,y) = \sum_{k=1}^{M} G_k(x,y) \tag{5}$$

5. PA signal and image processing

In PA imaging, ultrasonic signals are received in noise; thus, small objects become buried in noise. Hence, prior to image formulation, different signal processing techniques can be used to remove the noise, enhancing the signal-to-noise ratio. Filtering in addition to principal component analysis (PCA) are widely used techniques [15–24]. Also, multiresolution analysis, utilizing wavelet analysis and subband decomposition, of PA signals can be used to obtain images at different scales [17, 20–24]. Furthermore, post-image formulation; image processing methods can be used to enhance the visualization of the image. These methods include contrast enhancement, edge detection, segmentation, and pattern

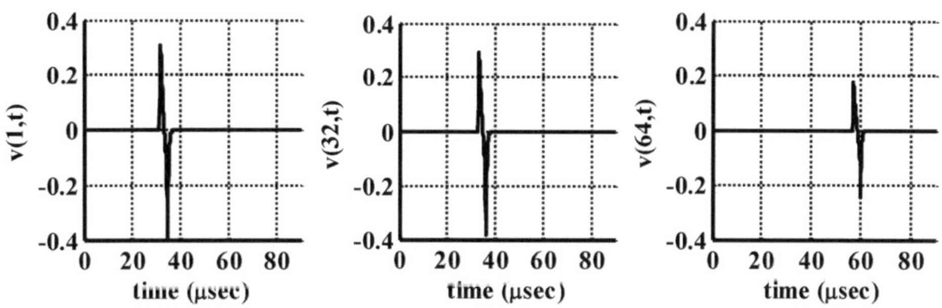

Figure 4.
Signals received at sensors number 1 (most top-left side), 32 and 64 (most top-right side) in the system geometry of ***Figure*** *3. The simulated tumor is a uniform circular disk of radius 0.25 cm centered at (−2, −2). The set of the signals are used for the construction of the image in* ***Figure*** *5.*

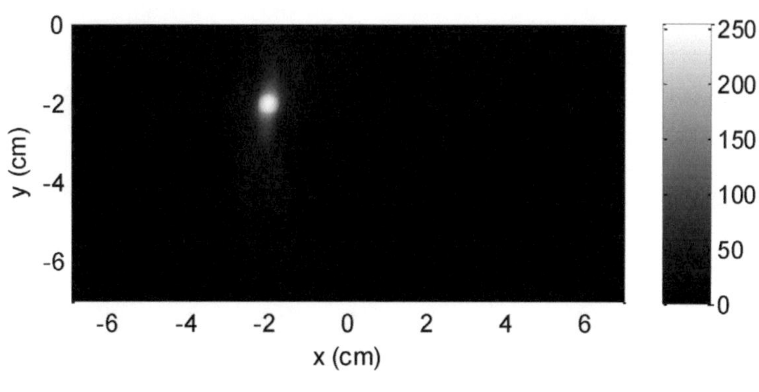

Figure 5.
Image formation by the back projection algorithm in (4) and (5). It is obvious that the tumor is extracted. Notice that the problem here is simplified, no noise is assumed, the light distribution is assumed uniform within the overall examined object, which is not practically a simple task.

recognition [4]. Deep learning is expected to play a good role in extracting more features PA imaging. In [25], deep learning is used to construct PA image from sparse data (**Figure 4**).

6. Conclusion

Simply saying, PA imaging utilizes the ultrasonic signal sent by the examined object to formulate an image for the distribution of the NIR light absorption coefficient within this object. This distribution differentiates abnormal tissues (malignant) from the normal one. Utilizing the absorption coefficients due to two different wavelengths provides information about the oxygen saturation. Hybrid modalities by combing PA image and pure ultrasound image offer imaging at cellular and molecular level. The challenges associated with the PA imaging modality are to deliver the examined object with a uniform distributed light independent of the depth and shape of this object and to reconstruct an online three-dimensional image for image-guided biomedical applications.

Author details

Reda R. Gharieb
Faculty of Engineering, Assiut University, Egypt

*Address all correspondence to: rrgharieb@gmail.com

References

[1] Oraevsky AA, Jacques SL, Tittel FK. Determination of tissue optical properties by piezoelectric detection of laser-induced stress waves. In: Proceedings of SPIE (The Society of Photo-Optical Instrumentation Engineers). Photons Plus Ultrasound: Imaging and Sensing, Vol. 1882, January 1993. 1993. pp. 86-101

[2] Shi J et al. High-resolution high-contrast mid-infrared imaging of fresh biological samples with ultraviolet-localized photoacoustic microscopy. Nature Photonics. 2019;**13**(9):609-615

[3] Lin L, Wang LV. Single-breath-hold photoacoustic computed tomography of the breast. Nature Communications. 2018;**9**(2352):1-9

[4] Wang B, Su JL, Amirian J, Litovsky SH, Smalling R, Emelianov S. Detection of lipid in atherosclerotic vessels using ultrasound-guided spectroscopic intravascular photoacoustic imaging. Optics Express. 2010;**18**(5):4889-4897

[5] Mallidi S, Luke GP, Emelianov S. Photoacoustic imaging in cancer detection, diagnosis, and treatment guidance. Trends in Biotechnology. 2011;**29**(5):213-221

[6] Ermilov S et al. Detection and noninvasive diagnostics of breast cancer with 2-color laser optoacoustic imaging system. In: Proceedings of SPIE. Photons Plus Ultrasound: Imaging and Sensing, Vol. 6437. 2007. pp. 1-11

[7] Mehrmohammadi M, Yoon SJ, Yeager D, Emelianov SY. Photoacoustic imaging for cancer detection and staging. Current Molecular Imaging. 2013;**2**(1):89-105

[8] Yaseen MA, et al. Hybrid optoacoustic and ultrasonic imaging system for detection of prostate malignancies. In: Proceedings of SPIE. Photons Plus Ultrasound: Imaging and Sensing. 2008

[9] Wang LV. Tutorial on photoacoustic microscopy and computed tomography. IEEE Journal of Selected Topics in Quantum Electronics. 2008;**14**(1):171-179

[10] Anastasio MA. Advancements in image reconstruction for photoacoustic computed tomography. In: Proceedings of SPIE. Photons Plus Ultrasound: Imaging and Sensing, Vol. 108780, March 2019. 2019

[11] Hariri A et al. The characterization of an economic and portable LED-based photoacoustic system to facilitate molecular imaging. Photoacoustics. 2018;**9**:10-20

[12] Wu Z et al. A microrobotic system guided by photoacoustic computed tomography for targeted navigation in intestines in vivo. Science robotics. 2019;**4**(32):1-11

[13] Xu Y, Wang LV, Ambartsoumian G, Kuchment P. Reconstructions in limited-view thermoacoustic tomography. Medical Physics. 2004;**31**(4):724-733

[14] Xu M, Wang LV. Universal back-projection algorithm for photoacoustic computed tomography. Physical Review E. 2005;**71**(1):016706-1:7

[15] Wang Y, Xing D, Zeng Y, Chen Q. Photoacoustic imaging with deconvolution algorithm. Physics in Medicine and Biology. 2004;**49**:3117-3124

[16] Manohar S, Dantuma M. Current and future trends in photoacoustic breast imaging. Photoacoustics. 2019;**16**(100134):1-27

[17] Zeng L, Xing D, Gu H, Yang D, Yang S, Xiang L. High antinoise photoacoustic tomography based on a

modified filtered backprojection algorithm with combination wavelet. Medical Physics. 2007;**34**(2):556-563

[18] Li PC, Wei CW, Sheu YL. Subband photoacoustic imaging for contrast improvement. Optics Express. 2008; **16**(25):20215-20226

[19] Gao F, Feng X, Zhang R, Liu S, Zheng Y. Adaptive photoacoustic sensing using matched filter. IEEE Sensors Letters. 2017;**1**(5):1-3

[20] Holan SH, Viator JA. Automated wavelet denoising of photoacoustic signals for circulating melanoma cell detection and burn image reconstruction. Physics in Medicine and Biology. 2008;**53**:N227-N236

[21] Zalev J, Kolios MC. Detecting abnormal vasculature from photoacoustic signals using wavelet-packet features. In: Proceedings of SPIE. Photons Plus Ultrasound: Imaging and Sensing, Vol. 7899. 2011. p. 78992M-1:15

[22] Zhou M, Xia H, Lan H, Duan T, Zhong H, Gao F. Wavelet de-noising method with adaptive threshold selection for photoacoustic tomography. In: Proceedings of the Annual International Conference of the IEEE Engineering in Medicine and Biology Society, July 2018. 2018. pp. 4796-4799

[23] Zhou M et al. A noise reduction method for photoacoustic imaging in vivo based on EMD and conditional mutual information. IEEE Photonics Journal. 2019;**11**(1):1-10

[24] Frikel J, Haltmeier M. Efficient regularization with wavelet sparsity constraints in photoacoustic tomography. Inverse Problems. 2018; **34**(2):1-28

[25] Antholzer S, Haltmeier M, Schwab J. Deep learning for photoacoustic tomography from sparse data. Inverse Problems in Science and Engineering. 2019;**27**(7):987-1005

Chapter 2

Photoacoustic Spectroscopy of Gaseous and Condensed Matter

Surya Narayan Thakur

Abstract

A brief historical account of photoacoustic (PA) effect is followed by a simple mathematical model for the generation of PA signals due to nonradiative transitions in atoms and molecules. Some experimental setups, with microphone and piezoelectric transducers, are described for recording PA spectra of gaseous, solid, and liquid samples. Applications of PA spectroscopy in the investigation of harmful chemicals are presented with illustrative examples. The principle of photoacoustic imaging (PAI) is discussed along with examples of molecular imaging of biological tissue and internal organs in small animals.

Keywords: photoacoustic effect, nonradiative decay, microphone, piezoelectric transducer, photoacoustic cell, aerosols, explosives, actinides, hyperspectral imaging, molecular imaging, photoacoustic microscopy (PAM), photoacoustic tomography (PAT)

1. Introduction

Photoacoustic imaging (PAI) is a novel method of obtaining spectral images of chemical constituents of a sample or a scene, to gain valuable insight into its structure and dynamics. It is based on the technique of photoacoustic spectroscopy (PAS) and covers the entire spectral range from the ultraviolet to the infrared. When light is incident on a sample, photons can be either absorbed, transmitted, or reflected, and the PAS technology focuses on the amount of absorption and its subsequent release as heat. PAS is an extremely sensitive detection technique as it can detect molecular concentrations below the parts per billion (ppb) level. This technique emerged from the discovery of the photoacoustic effect by Graham Bell in 1880 during his attempt to transmit sound over a beam of sunlight [1]. However, it remained dormant for almost a century until the advent of tunable lasers in the 1970s and was successfully used by Kreuzer and Patel for the trace detection of atmospheric pollutants [2].

1.1 Photophone and the spectrophone

Bell succeeded in wireless audio communication about two decades before the radio transmission. He used the newly discovered selenium cell in the receiver in view of selenium's property to react to modulated intensity of sunlight incident on it, as the resistance of selenium crystal depends on the incident light. A flexible mirror was attached at the speaking end of the photophone that created slight deviation of the beam of light reaching the receiver end. This led to variation of intensity at the selenium receiver, which acted like an optical version of the electric coil in the telephone receiver, converting the intensity modulated light back into sound. Bell performed many experiments and observed that sound waves were also produced directly from a solid sample when exposed to a periodically modulated beam of sunlight as illustrated in **Figure 1**. A hearing tube, whose other end was tightly attached to the open end of a transparent glass test tube with sample placed at its closed end, could be used as a photophone. When a beam of sunlight focused on the sample was rapidly interrupted with a rotating slotted wheel at an audible frequency, the intensity of sound in the hearing tube was dependent on the type of material. The loudest sound was heard when the sample was carbon black, leading to the conclusion that photoacoustic effect was caused by the absorbed light energy which subsequently heats the sample.

During Bell's visit to England in 1880, John Tyndall performed the photoacoustic experiment in gases, and although the photoacoustic effect was confirmed, Tyndall was of the view that it was caused mainly by the radiant heat [3]. Bell was driven by rare intellectual curiosity to learn, and it led him to invent the spectrophone to find out the wavelengths that were more efficient for the radiant heat [4]. For this purpose he converted a prism spectroscope into a spectrophone by replacing the eyepiece of the telescope with a hearing tube in which a thin wire mesh coated with lampblack was fitted in the position of the cross wires (see **Figure 2**). When the incident sunlight was interrupted by a mechanical chopper, the hearing tube produced sound whose frequency was equal to the periodic intermittence of light. The loudness of sound, however, varied in accordance with the intensity of the solar spectrum, being maximum in the green-yellow region and decreasing at both the red and violet ends, and observations were made by fixing the position of the telescope in different spectral regions of the solar spectrum. These observations confirmed the fact that the photoacoustic effect is due to optical absorption, since lampblack totally absorbs light at each wavelength. On the basis of his observations, Bell made the prophetic statement about the great importance of photoacoustic spectroscopy in the infrared.

The interaction between light and matter giving rise to photoacoustic effect has three distinct features. (1) The absorbed energy of optical radiation is converted into heat. (2) At the site of optical absorption, there is a temporal rise and fall of

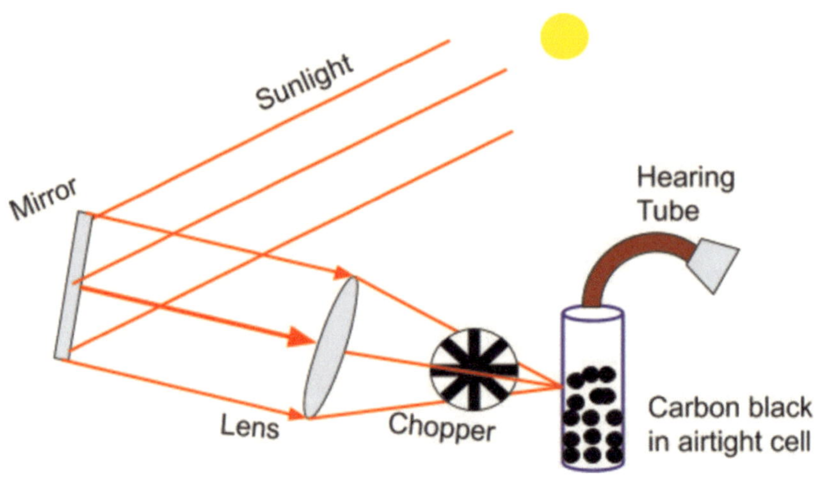

Figure 1.
Illustration of photoacoustic effect with periodically chopped light incident on carbon black sample cell fitted with a hearing tube.

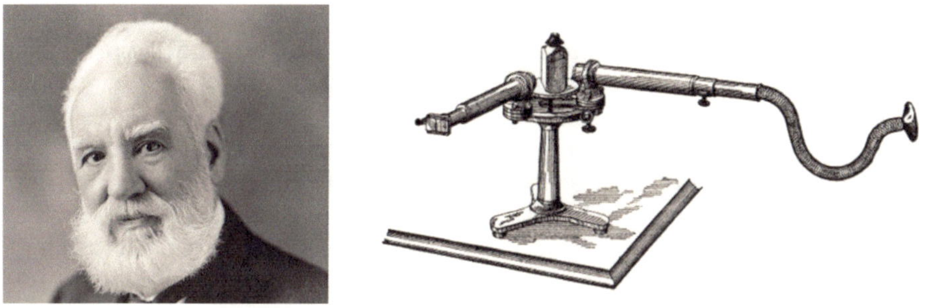

Figure 2.
Alexander Graham Bell and his spectrophone.

temperature. (3) The expansion and contraction following these temperature changes lead to periodic pressure variation to generate sound.

2. Nonradiative transitions and PA signal generation

The heat generation following optical absorption is caused by internal motions in molecules or those of the matrix in which atoms are imbedded in condensed matter. In the quantum mechanical description, an excited molecular state reached by the optical absorption has two channels of relaxation. The radiative decay leads to optical emission, whereas the nonradiative decay causes heating. Thus a molecule, optically excited to a vibronic or a rovibrational state, loses a part of its excitation energy as heat leading to the photoacoustic signal. The photoacoustic spectrum is similar to the absorption spectrum, but its intensity at the exciting wavelength is proportional to the product of the absorption coefficient and the probability of nonradiative decay of the excited state.

Figure 3 shows the energy level diagram, in the Born-Oppenheimer approximation, of a typical organic molecule with an even number of electrons where total internal energy $E = E_e + E_v + E_r$. The rotational energy levels are not shown for the sake of simplicity, and only one of the 3N-6 (or 3N-5) normal modes for a nonlinear

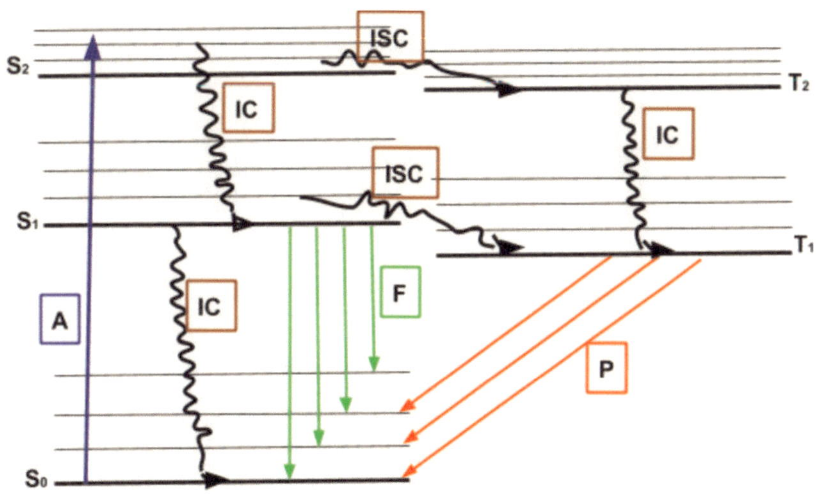

Figure 3.
Typical energy level diagram of an organic molecule where radiative processes A, F, and P stand for absorption, fluorescence, and phosphorescence, respectively. The nonradiative processes IC and ISC refer to internal conversion (IC) and intersystem crossing (ISC), respectively.

(or linear) molecule with N atoms is depicted for the various electronic states. In a large molecule like benzene, nonradiative decay from excited states occurs even in the vapor phase. Michael Kasha pioneered the investigations on nonradiative transitions, and it was found that radiative emission occurs only from the lowest excited state of any multiplicity [5]. The concept of triplet state (T) was first given by Lewis and Kasha [6] to explain the origin of phosphorescence emission in organic or inorganic molecules. Internal conversion (IC) refers to the nonradiative transfer of a molecule from an excited vibronic state to a lower vibronic state of the same multiplicity, whereas intersystem crossing (ISC) corresponds to nonradiative transition between a singlet (S) and a triplet (T) state. Lewis and Kasha formulated a fundamental law concerning fluorescence and phosphorescence through intersystem crossing [6].

2.1 Generation of photoacoustic signal in gases

The rate of radiative transition $\mathbf{r}_{ij}$ between two states of a molecule E_i and $E_j(<E_i)$ in terms of Einstein coefficient of stimulated (B_{ij}) and spontaneous (A_{ij}) emission is of the form

$$\mathbf{r}_{ij} = \rho(\nu)B_{ij} + A_{ij} \tag{1}$$

where $B_{ij} = B_{ji}$ and $\rho(\nu)$ is the spectral energy density at the frequency ν of the transition. For a two-level system of energy E_0 and E_1 shown in **Figure 4**, we have $B_{01} = B_{10}$ and $A_{01} = 0$, since spontaneous emission from a state of lower energy to one of higher energies does not occur. If c_{ij} represents the probability of a nonradiative (collisional) transition between a pair of states, we assume the collisional excitation from E_0 to E_1 to be very small so that $c_{01} = 0$.

Suppose the number density of molecules in the ground and excited states of **Figure 4** is n_0 and n_1, respectively. In view of the above description of the rate of change, the number density in the excited state by radiative and nonradiative transitions is given by

$$\begin{aligned} dn_1/dt &= (r_{01} + c_{01})n_0 - (r_{10} + c_{10})n_1 = \rho(\nu)B_{01}n_0 - (\rho(\nu)B_{10} + A_{10} + c_{10})n_1 \\ &= \rho(\nu)B_{01}(n_0 - n_1) - (A_{10} + c_{10})n_1 \end{aligned} \tag{2}$$

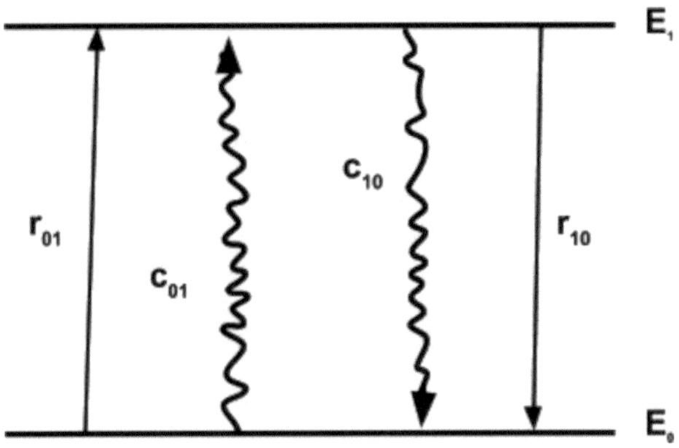

Figure 4.
Two-level molecular model showing the rate of radiative (r_{ij}) and nonradiative (c_{ij}) transitions between energy states E_0 and E_1.

We define the radiative and nonradiative lifetimes to be $\tau r = 1/A_{10}$ and $\tau c = 1/c_{10}$, respectively, and the total lifetime $\tau(1/\tau = 1/\tau r + 1/\tau c)$, to get the following expression for the rate of change of excited state population:

$$\begin{aligned} dn_1/dt &= \rho(\nu)B_{01}(n_0 - n_1) - (1/\tau r + 1/\tau c)n_1 \\ &= \rho(\nu)B_{01}(n_0 - n_1) - n_1/\tau \end{aligned} \tag{3}$$

In a similar manner, we find the following expression for the rate of change of ground state population density:

$$dn_0/dt = \rho(\nu)B_{01}(n_1 - n_0) + n_1/\tau \tag{4}$$

Hence from Eqs. (3) and (4), we get

$$dn_1/dt - dn_0/dt = 2\rho(\nu)B_{01}(n_0 - n_1) - 2n_1/\tau \tag{5}$$

In a photoacoustic experiment, we assume the incident light intensity I to vary slowly so that we may consider the upper and lower state population density changes to be an adiabatic interchange. Under this approximation we can set the left-hand side of Eq. (5) to zero. Since the total molecular density for the two-level system is $N = n_1 + n_0$, Eq. (5) takes the following form:

$$0 = 2\rho(\nu)B_{01}(n_0 - N + n_0) - 2(N - n_0)/\tau = 4\rho(\nu)B_{01}n_0 - 2N(\rho(\nu)B_{01} + 1/\tau) + 2n_0/\tau$$

$$\text{Hence} \quad n_0 = N[\rho(\nu)B_{01} + 1/\tau]/(2\rho(\nu)B_{01} + 1/\tau) \tag{6}$$

$$\text{Similarly} \; n_1 = N\rho(\nu)B_{01}/(2\rho(\nu)B_{01} + 1/\tau) \tag{7}$$

The spectral radiant energy density is directly proportional to the intensity I of the light source, and we define a constant $B = \rho(\nu)B_{01}/I$ so that Eq. (7) takes the following form:

$$n_1 = NBI/(2BI + 1/\tau) \tag{8}$$

Mechanical chopping of the light source at a frequency ω can be expressed as the periodic "on" and "off" of the intensity by $I = I_0 \cos\omega t$ for simplicity, and we get the following expression for the excited state population density:

$$n_1 = NBI_0\cos\omega t/(2B1_0\cos\omega t + 1/\tau) \tag{9}$$

If the gas is very weakly absorbing, we assume that most of the molecules are in state E_0 or $n_0 >> n_1$. Hence from Eqs. (6) and (7), $BI < < 1/\tau$ so that Eq. (9) takes the following form:

$$n_1 = \tau NBI_0\cos\omega t \tag{10}$$

For a molecular gas, its kinetic energy (K) has to be considered to get the total internal energy density U of the two-level system leading to

$$U = n_1E_1 + K \tag{11}$$

The rate of change of energy is given by

$$dU/dt = (dn_1/dt)E_1 + dK/dt = [\rho(\nu)B_{01}(n_0 - n_1) - (A_{10} + c_{10})n_1]E_1 + dK/dt \tag{12}$$

This change of energy is equal to the difference between the absorbed and radiated optical energy so that

$$\begin{aligned} dU/dt &= (r_{01}n_0 - r_{10}n_1)E_1 \\ &= [\rho(\nu)B_{01}(n_0 - n_1) - A_{10}n_1]E_1 \end{aligned} \tag{13}$$

We know that $A_{01} = 0$ and $c_{01} = 0$; hence from Eqs. (12) and (13), we get the following relation:

$$dK/dt = c_{10}n_1E_1 \tag{14}$$

Since the volume of the photoacoustic cell does not change in the experiment, and thermodynamic evaluation of the change in kinetic energy at constant volume gives

$$\begin{aligned} dK &= (\delta K/\delta T)_V dT + (\delta K/\delta V)_T dV \\ &= (\delta K/\delta T)_V dT \end{aligned} \tag{15}$$

Since the specific heat capacity of gas $C_V = (\delta K/\delta T)_V$, by integrating Eq. (15) we get

$$K = C_V T + f(V) \tag{16}$$

where $f(V)$ is a constant of integration.

The pressure of the ideal gas $P = Nk_BT$, where N is the number density of molecules, k_B is the Boltzmann constant, and T is the temperature. Thus substituting for T from Eq. (16) in the expression for P, we get

$$P = Nk_B[K - f(V)]/C_V \tag{17}$$

Since $f(V)$ is constant and the pressure wave (sound) is given by $\delta P/\delta t$, we get the following relation from Eq. (17) in the light of Eq. (14):

$$\begin{aligned} \delta P/\delta t &= (Nk_B/C_V)dK/dt \\ &= (Nk_B/C_V)c_{10}n_1E_1 \end{aligned} \tag{18}$$

Substituting for n_1 from Eq. (10) and using the relation $c_{10} = 1/\tau c$, Eq. (18) reduces to

$$\delta P/\delta t = \left(k_BN^2E_1/C_V\right)(\tau/\tau c)BI_0\cos\omega t \tag{19}$$

Integrating Eq. (19) we get the following expression for photoacoustic signal, which is detected by a sensitive microphone:

$$P(t) = \left(k_BN^2E_1/C_V\right)(\tau/\tau c)BI_0\sin\omega t \tag{20}$$

2.2 Generation of photoacoustic signal in solids

Let us consider a cylindrical photoacoustic cell shown in **Figure 5** where the light-absorbing solid sample is surrounded by optically transparent gas (air) on the front side and the backing material is a poor conductor of heat. The absorption of light, of a particular wavelength in the sample, generates heat by nonradiative

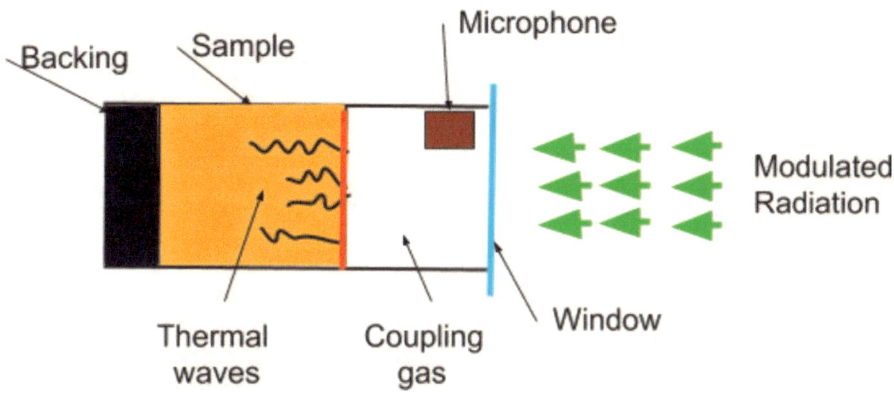

Figure 5.
Photoacoustic cell for a solid sample. Thermal waves originate from the point of absorption and travel toward the solid–gas interface to periodically heat a thin layer of gas (shown in red).

transitions. The acoustic signal produced in the coupling gas is due to periodic heat flow from the sample. Rosencwaig-Gersho model for PA signal generation depends on the optical as well as thermal properties of the solid sample [7]. The periodic heat flow from the solid sample heats a thin layer of coupling gas whose thickness depends on the frequency of chopping of the incident light (about 1 mm at 100 Hz). This layer (shown in red in **Figure 5**) may be imagined as a vibratory gas piston that creates the acoustic signal detected by the microphone. The intensity of light of wavelength λ transmitted through thickness "x" of the solid sample of absorption coefficient β is given by

$$I(\lambda) = I_0\, e^{-\beta x} \tag{21}$$

where I_0 is the incident intensity for wavelength λ, and if ω is the modulation frequency of the incident radiation, the temporal variation of I_0 is given by

$$I_0(t) = (1/2)\, I_0 (1 + \cos\omega t) \tag{22}$$

The temperature variation in the gas dies out within a thickness of 2πμ from the surface of the solid sample where μ is thermal diffusion length of the gas.

3. Experimental methods in photoacoustic spectroscopy (PAS)

The temperature and pressure changes involved in the process of PA signal generation are extremely small, typically a micro- to millidegree and nano- to microbar, respectively. This was the reason that the field of PAS remained dormant till the advent of tunable laser sources and of sensitive audio detectors. A piezo-electric transducer or a sensitive microphone serves as the acoustic detector in photoacoustic cells used during the laboratory experiments. Recent developments in the field of miniaturization and the related progress in computer software have made it possible to use tiny quartz-based acoustic detection devices like cantilever and crystal tuning forks. The photoacoustic spectrometer using miniature lasers fitted with these novel detectors can be easily packed in a small box and can be used in the laboratory or in the field for standoff detection of hazardous materials. The heat generated by the photoacoustic effect produces density changes caused by temperature fluctuations in liquid and gaseous samples. In such cases, photoacoustic spectroscopy is carried out by detection of thermal lens formation using a probe laser. In the present case, however, we will confine to the detection of acoustic vibrations in the analysis of PA spectra.

3.1 Photoacoustic cells for gaseous samples

One of the simplest PA cells for measurements on gases and vapors is made from a Pyrex tube fitted with quartz windows at the two ends. The length of the tube and location of the microphone are chosen to maximize the PA signal using the resonance of sound generated by the modulated light beam. The design of such a cell made from 62.5-cm-long Pyrex tube of 2.5 cm diameter and fitted with quartz windows at Brewster's angle is shown in **Figure 6(a)**. The cell was resonant at 335 Hz with maximum signal in the middle port for the microphone, and it was resonant at 669 Hz, rendering maximum signal at two ports symmetrically located on either side of the middle port. The microphone ports correspond to the positions of three possible antinodes of the stationary acoustic waves formed at the two resonant frequencies. Acoustic isolation was achieved by putting the cell in a wooden box filled with sand. Two of the ports, not in use during the measurement, were sealed using O-rings and flat Teflon disks. PA measurements were carried out with iodine vapor at room temperature in the presence of air at atmospheric pressure using 20 milliwatt Argon laser light at 514.5 nm. The chopping frequency of the

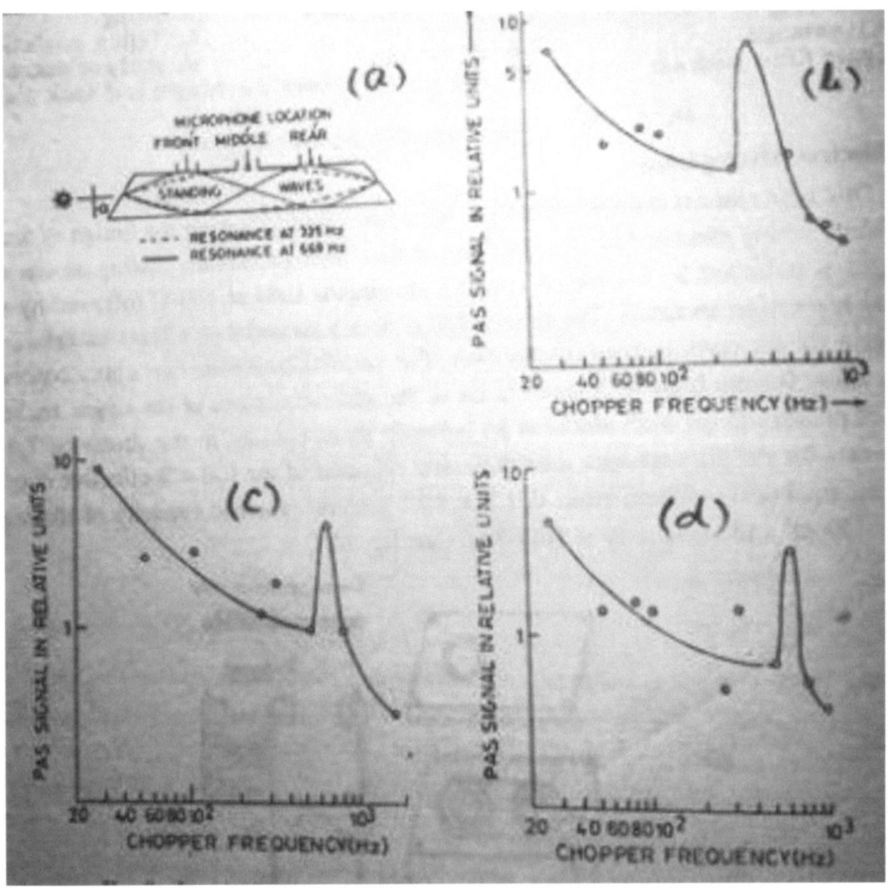

Figure 6.
Longitudinally resonant PA cell with three ports for microphone (a). PA signal for resonance at 335 Hz (b) and those for resonance at 669 Hz (c) and (d).

laser light was varied between 27 and 1000 Hz, and the resulting PA signals are shown in **Figure 6(b–d)** [8].

The photoacoustic spectrum of iodine vapor recorded using a Nd:YAG laser pumped tunable dye laser in the presence of atmospheric air and at 15 Torr is shown in **Figure** 7. Dye laser pulses used for the measurements were of 7 ns duration, 0.05 nm bandwidth, and 2mJ/pulse energy at a pulse repetition rate of 10 Hz. It is seen from the spectrum recorded with low air pressure in **Figure 7b** (upper half) that the relative intensity of vibronic bands monotonically increases in going up to the dissociation limit where I_2 molecule dissociates into a ground-state $(^2P_{3/2})$

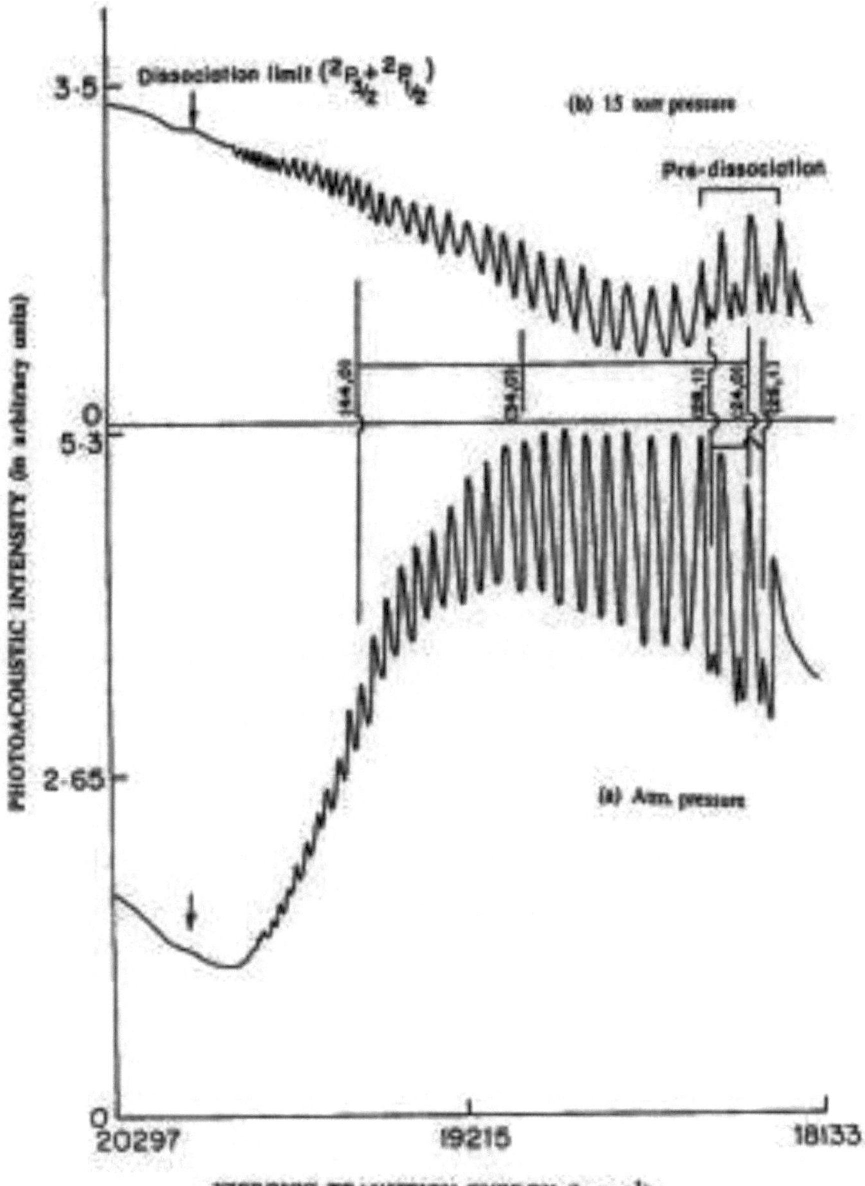

Figure 7.
Photoacoustic spectra of I_2 vapor in the wavelength region 492–552 nm (20,300–18,100 cm^{-1}). The little downward arrow indicates the wavelength of exciting radiation that dissociates I_2 into two iodine atoms (20,043 cm^{-1}).

iodine atom and an excited-state $\left({}^2P_{1/2}\right)$ iodine atom. In **Figure 7a** for the spectrum recorded at atmospheric air, the relative intensity of vibronic bands monotonically decreases after maximum intensity at 19,225 cm^{-1} This reduction in intensity of the PA spectrum has been explained in terms of energy transfer from excited $\left({}^2P_{1/2}\right)$ iodine atoms to O_2 molecules following the dissociation of I_2 [9].

To make photoacoustic measurements on a flowing gas sample, such as in the case of pollution monitoring, one needs a different type of acoustic resonant PA cell as schematically illustrated in **Figure 8**. The U-shaped cell has a total length $L = \lambda$, corresponding to the acoustic wavelength generated by the periodically modulated laser. The horizontal section of the tube is of length $\lambda/2$ and two vertical sections at each end of length $\lambda/4$. At the resonance frequency, a stationary wave is formed with antinodes at the two ends as well as at the center of the horizontal section and nodes at the two bent corners (see **Figure 8**). The laser as well as gas inlet and outlet, in the PA cell, is located near the nodes, which form regions of low pressure, of the stationary acoustic wave to prevent pressure fluctuations in PA signal at the antinodes. A microphone is fixed near the open end of one of the vertical arms, to detect the PA signal. A piezoelectric disk, fixed at the end of the other vertical arm, is used for detection of resonance frequency of the PA cell and the resonator quality factor to calibrate the system. There is a phase difference of 180° between high pressure points of the stationary acoustic wave at the center of the horizontal section, where heat is generated, and the two ends where microphone and piezo-electric signals are measured. The flow rate of the sample in the PA cell is controlled by a very narrow orifice at the outlet end, such that the pressure inside the PA cell remains steady. The special narrow hole causes a pressure drop of at least 50% on the outlet side compared to pressure inside the PA cell, while the airflow velocity through the orifice reaches the speed of sound. Since the air traveling through the orifice is moving out at the speed of sound, any sound from the pump traveling toward the PA cell, also at the same speed, meets a barrier in air, moving in the opposite direction, and it goes no further. This arrangement blocks the noise from the pump entering the PA cell and cluttering the PA signal.

The airborne particulate matter (aerosols) remains suspended in air inside the PA cell shown in **Figure 8** and has enough time to absorb radiation from the tunable laser beam. The absorbed optical energy is transferred as heat to the surrounding air, before the next laser pulse arrives, to build the stationary pressure wave in the

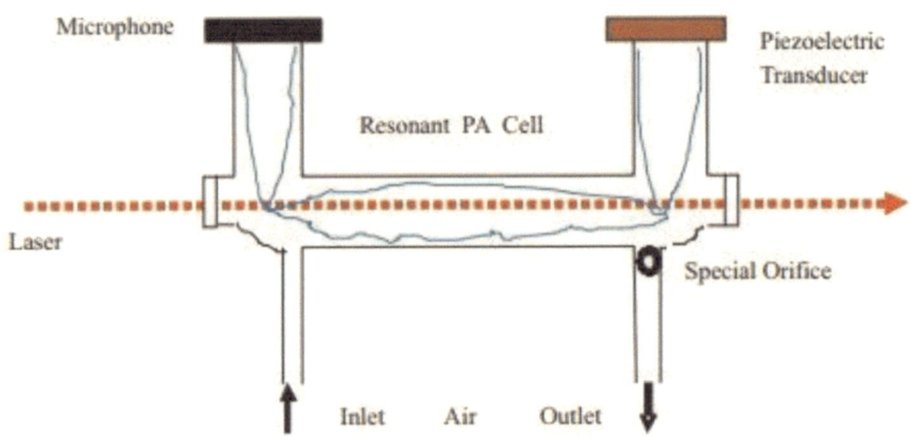

Figure 8.
The U-shaped resonant PA cell for detection of PA signal in a flowing air sample with special orifice for controlling the pressure inside the cell.

PA cell. An instrument of this type can directly measure light absorption by aerosols over the entire range of sunlight entering the atmosphere. This type of PA cell has been used with a single laser as well as with two lasers of different wavelengths impacting the same gaseous sample [10, 11].

3.2 Photoacoustic cells for solid samples

PA cells fitted with microphone, for recording PA spectra of solid samples, have been routinely used in the laboratory for almost four decades. One of the important aspects of homemade cells is to choose a material for effective shielding from extraneous sound. The design of a nonresonant PA cell is schematically shown in **Figure 9**. The main body of the cell has been constructed from a single block of aluminum with a cavity made from the bottom side for fixing the microphone along with its preamplifier. A cavity is made on the top of the block to put the sample cuvette whose open end is in the same horizontal plane as the microphone surface. A flat aluminum plate with double quartz windows in front of the sample cuvette is tightly fixed at the top of the main body with a very thin, suitably cut rubber sheet to make the chamber airtight. The thickness of the air duct connecting the sample and microphone is about 1 mm, and its total volume is less than 1 cc. The exterior dimensions of the stainless steel sample cuvettes are identical so as to tightly fit into their designated cavity. The sample cuvettes are, however, of varying depths to make measurements on powder samples of different thicknesses. Carbon black is used as the standard sample for recording the power spectrum of the excitation source of light to normalize the PA signals of the sample under investigation.

The PA spectra of powder samples of RDX and TNT recorded with a PA cell of the above type are shown in **Figure 10**. The source of excitation used in these experiments was a rotational line tunable cw CO_2 laser, and the window, in front of the sample in the PA cell, was made of flat ZnSe plate. The PA signals were measured manually from the lock-in amplifier and normalized using the power meter reading for each laser line. Microgram quantities of powder samples were used in the spectral measurements which exhibit the characteristic vibrational

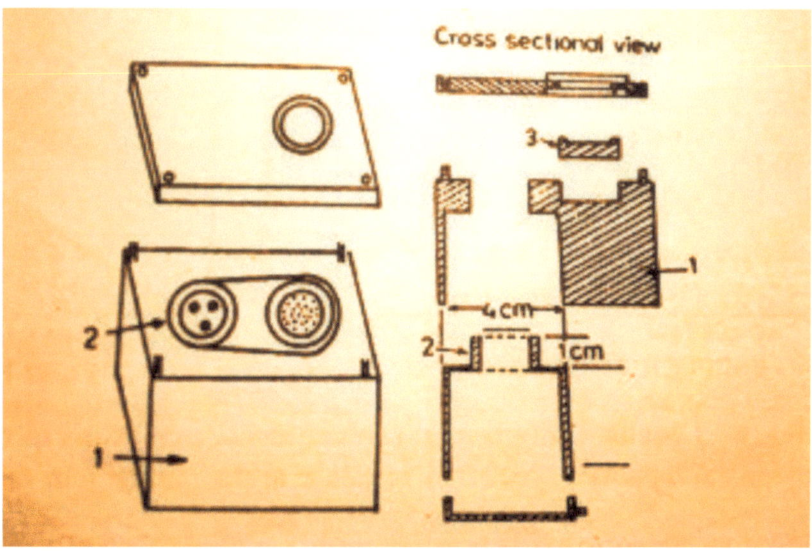

Figure 9.
Design of nonresonant PA cell. (1) the main body of aluminum, (2) microphone and preamplifier chamber, and (3) sample cuvette.

bands of RDX and TNT [12]. The solid circles in the upper half of **Figure 10** represent the normalized PA signals for the pure explosive powders in the rotational line tunable CO_2 laser wavelength range from 9.6 to 10.6 μm. The samples were diluted by uniformly mixing 10 micrograms of each explosive with 8 grams of SiO_2, which does not have any absorption in this region. Thus, the concentrations of the two explosives correspond to a few ppm, exhibiting four persistent bands of RDX and five of TNT with appreciable intensity as shown in the lower part of **Figure 10**.

3.3 PA spectroscopy of contaminated water

PA instrumentation for detection of liquid samples is somewhat complicated. A schematic diagram of the experimental setup for detection of harmful and dangerous pollutants in water is shown in **Figure 11**. The tunable dye laser beam, pumped by an excimer laser or a Nd:YAG laser, is focused into a 600 micron core multimode optical fiber for investigation on remotely located samples. The polluted water is kept in a quartz cuvette which is acoustically coupled to a piezoelectric transducer. The light exiting from the optical fiber is collimated into the quartz cuvette by means of a 10X microscope objective. The pulse to pulse energy fluctuation, in the light exiting from the cuvette, is monitored by a photodiode assembly, to ratio the PA spectra with the dye laser power profile. The PA signal and the photodiode signal are fed to two different gates of the boxcar, and the normalized spectrum is recorded by the computer by scanning the dye laser wavelength.
The quartz cuvette with its piezoelectric transducer, the microscope objective, and the photodiode are fixed on a small portable platform which can be easily positioned within the glovebox for handling radioactive samples. Klenze et al. [13] have determined ultralow concentrations of Am^{3+} and Pu^{4+} in aqueous solutions with excimer laser-pumped dye laser, while Nd : YAG-pumped dye laser system was used by Russo et al. [14] to record PA spectra of aqueous solutions containing different concentrations of Pr^{3+} and Am^{3+}. Kim [15] has investigated the problem of actinide colloid generation in groundwater which plays a critical role in geochemical interaction and migration of actinides.

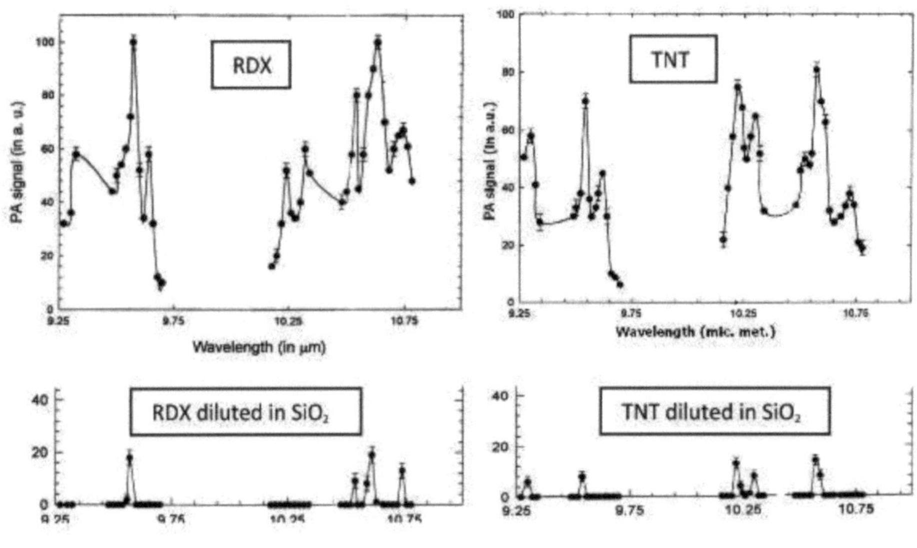

Figure 10.
CO_2 laser-excited PA spectra of RDX and TNT powders. The persistent vibrational bands of the two molecules are seen for highly diluted samples in the lower half of the figure.

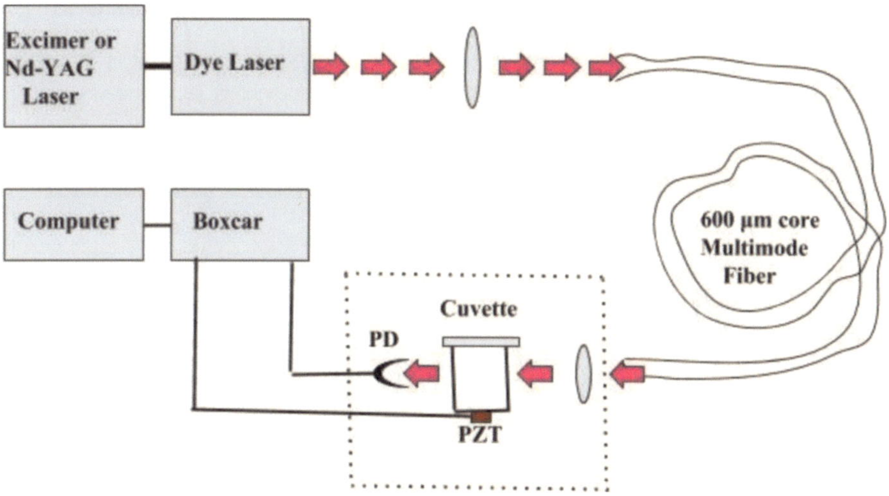

Figure 11.
Schematic PA spectroscopy system for trace detection of chemical species in polluted water.

3.4 Quartz tuning fork for PA detection

In a PA cell, the acoustic energy is accumulated in a resonant cavity, but the principle of PA detection by a quartz tuning fork (QTF) involves the accumulation of the acoustic energy in a sharply resonant acoustic transducer [16–18]. Crystal quartz is an easy material for such a transducer because of its low loss piezoelectric property, and QTFs can be designed to resonate at any frequency between 4 Hz and 200 kHz. The widely used QTF, manufactured for use in electronic clocks as frequency standard, resonates at 32,768 Hz in vacuum. The detection of PA signal by QTF is based on the piezoelectric effect produced by acoustic waves at the resonant frequency that mechanically bend its prongs. In the PA spectroscopy of gaseous sample, the interaction between the modulated laser beam and a trace gas generates acoustic waves that mechanically bend the QTF prongs. The voltage produced by the piezoelectricity generates the PA signal.

When a laser beam is focused at the center between the two prongs of the QTF placed in a gaseous sample, the absorbed optical energy converted into heat generates a weak acoustic pressure wave. When the laser beam is modulated at half the QTF resonant frequency (f), the pressure wave makes the two prongs move apart two times during each acoustic cycle. In this situation the QTF detects sound oscillations at the second harmonic of the modulation frequency due to two absorption events during each modulation period. The laser light is modulated at "$f/2$" and PA signal demodulated by lock-in amplifier at "f" as shown in **Figure 12**. The PA spectrum is recorded by varying the wavelength of the tunable laser.

The use of QTF for solid phase PA detection in the laboratory requires a very thin film of the molecular sample to be adsorbed on the outer surface of one of the prongs. The absorbed laser light heats the sample, generating an acoustic wave at the prong's surface interface with air. When the frequency of repetition of the incident laser pulse coincides with the mechanical resonant frequency of the QTF, the localized pressure variation sets the latter into vibration. The amplitude of this vibration and the resulting piezoelectric voltage are proportional to the amount of heat produced by optical absorption at the surface.

An experimental arrangement using the above concept is schematically shown in **Figure 13** using a quantum cascade laser (QCL). The large wavelength coverage in

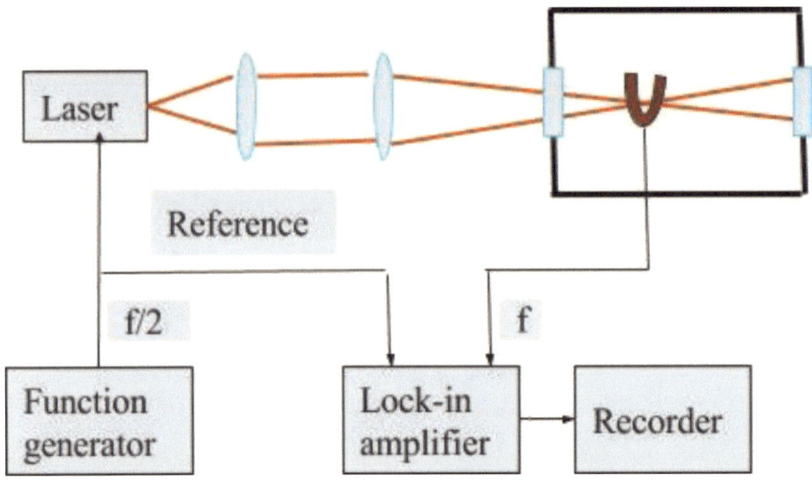

Figure 12.
Experimental setup for gas phase PA spectroscopy with QTF detector. The excitation diode laser source is currently modulated at half of the QTF resonant frequency (f).

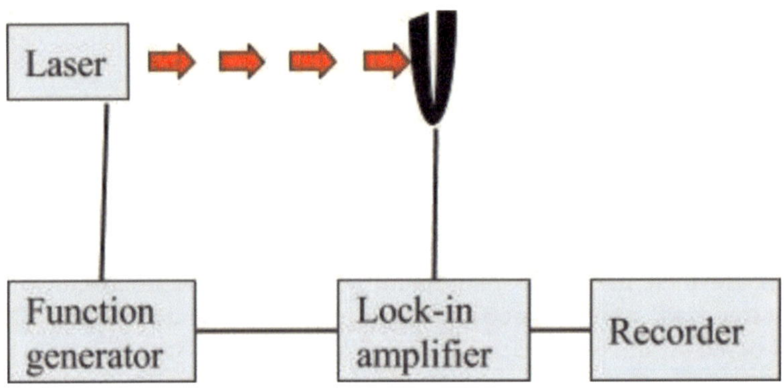

Figure 13.
Schematic experimental arrangement for solid phase PA spectroscopy using QTF detector with sample adsorbed on one of the QTF prongs. The function generator controls the pulse repetition rate of the quantum cascade laser (QCL) to be in resonance with the symmetric vibration of the QTF.

the mid-IR region combined with narrow linewidth and powering up to tens of mW of QCL proves ideal for trace detection of molecules [19]. This system can be employed in trace detection of several molecular species adsorbed on the surface of the prong. A variant of this system has been used in detection of adsorbed chemicals on surfaces remotely located from the laser [20].

4. Photoacoustic detection of harmful chemicals

PA spectroscopy has been widely used in chemical sensing applications in environmental science and medical diagnostics. It is useful in rapid detection of illicit drugs, nerve agents, and hazardous biological materials. In a typical hospital environment, there is a need for evaluation of anesthetic gaseous components. Although hospital staff are exposed to much lower anesthetic concentration than the patients, this exposure extends over many years. Under inadequate hygiene conditions, people working in hospitals or factories often complain of headaches and fatigue due to traces of harmful gases in the environment. Illicit drug trafficking poses many challenges for detection of dangerous chemicals that threaten life and property. In the following sections, we will present examples of point detection as well as standoff detection of chemical compounds using PA spectroscopy.

4.1 PA spectroscopy of ethylene

Ethylene (C_2H_4) is a well-known emission fingerprint from vehicle exhausts and its reaction with nitric oxides under solar UV radiation produces ozone. Altuzar et al. [21] have used a CO_2 laser with PA cell located inside the laser cavity to profit from the intracavity laser power. A mechanical chopper modulated the intensity of the laser beam at a frequency that matches with the acoustic resonant frequency of the PA cell. The 10P14 line of CO_2 laser is in exact resonance with the Q branch of the ν_7 band of ethylene. Air samples from different locations in Mexico city were collected from 6 to 9 am in the morning and from 12 to 15 pm in the afternoon. The raw samples, collected in stainless steel vessels, went through the process of removing CO_2 and water vapor before being transferred into the PA cell. The results of ethylene concentration analysis exhibit higher values in samples collected in the morning and lower values for the afternoon samples. The increased concentration of ethylene is attributed to emission from vehicles when the morning traffic is heavy. The decrease of ethylene concentration in the afternoon samples has two possible causes. Firstly, there is expansion in the thickness of atmospheric air, due to heating by sunlight leading to dilution of ethylene. Secondly, ethylene reacts with nitrogen oxides in the presence of solar UV radiation to produce ozone and other oxidants, causing a decrease in its concentration. It is to be emphasized that PA technique has the advantage of performing online measurements with excellent time resolution.

4.2 PA spectroscopy of gases emanating from the human body

The smells emanating from various parts of the body are unique to an individual, made up of specific chemical compounds that vary depending on age, diet, metabolism, and health. Near-IR diode laser at 1.53 μm has been used in PA detection of ammonia concentrations in ppmV and ppbV range by several workers [18, 22]. Although NH_3 is biosynthesized through normal amino acid metabolism, it is toxic in high concentrations, and its amount in blood gets elevated due to liver dysfunction. Compact quantum cascade lasers (QCL) may prove to be an important mid-IR source for portable ammonia sensors.

Trace level detection of nitric oxide has many applications in medicine, biology, and environmental science. CO laser was the first to be used by Kreuzer and Patel [2] for PA detection of NO concentrations of 0.01 ppmV. Since its first detection in exhaled air [23], NO has been found to be a sensitive marker for asthmatic airway inflammation [24]. A QCL-based PA cell has been developed by Elia et al. [25], while Spagnolo et al. have reported a minimum NO concentration limit of 15 ppbV [26].

A CO_2 laser has been used as the source of PA excitation by Harren et al. [27] to detect the emission of ethylene from the human breath and skin under exposure to UV radiation. In this experiment a small amount of exhaled air was cleaned for its content of CO_2, water vapor, and other spectroscopically interfering gases, before introducing it into the PA cell. The 10P14 emission line of CO_2 laser was used to detect a lower limit of 6 ppbV ethylene in nitrogen.

4.3 PA spectroscopy of dangerous drugs

Morphine is the prototype narcotic drug, and it is the standard against which all other opioids are tested. An acetylated form of morphine, almost two times more

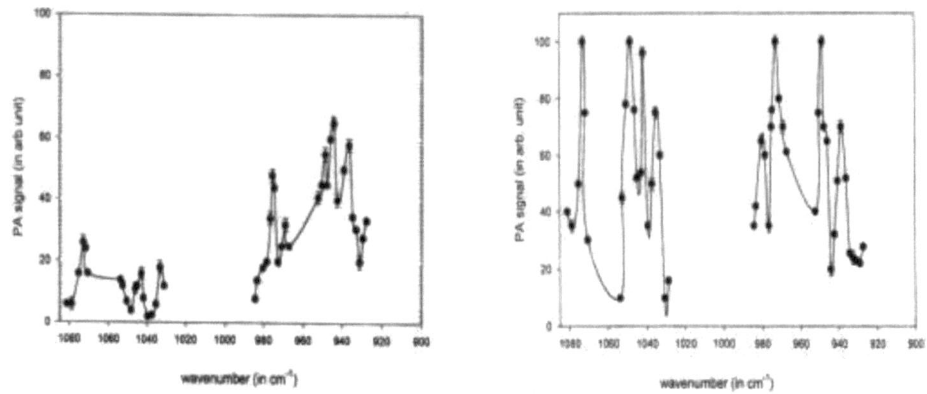

Figure 14.
CO_2 laser-excited PA spectra of powders of heroin (on the left) and morphine (on right).

potent than morphine itself, is known as heroin. Animal and human studies and clinical experience back up the contention that morphine is one of the most euphoric drugs on earth. Both morphine and heroin are used for pain medication, but both are addictive and identified as illegal drugs.

Microgram quantities of powders of morphine and heroin were used in a PA cell shown in **Figure 9** and fitted with ZnSe window. Rotational line tunable CO_2 laser was used as the excitation source for recording the PA spectra of the two compounds shown in **Figure 14**. The black dots in the figures represent the manually measured PA signals from the lock-in amplifier normalized by laser power at each wavelength. The spectral peaks correspond to the vibrational bands of morphine and heroin in the wavelength range 9.6 to 10.6 micron [28].

5. Photoacoustic imaging

Photoacoustic imaging is an emerging technique that combines the high resolution of light and deep imaging capability of ultrasound. It is similar to hyperspectral imaging except for the fact that optical sensors are replaced by ultrasonic detectors that convert the sound waves into images. It has many applications as a noninvasive technique in medicine to produce molecular images of internal organs. It is based on the rapid production of heat, when the optical energy from a nanosecond laser pulse is absorbed in the tissue, causing thermal expansion and the generation of ultrasonic waves. The processes involved in the image formation are schematically illustrated in **Figure 15**, which show that it is a hybrid technique making use of optical absorption and ultrasonic wave propagation.

There are two basic conditions for efficient generation of the PA signal for imaging. The condition of "thermal confinement" requires the laser pulse duration τ_p to be shorter than temporal duration τ_{th} of thermal diffusion from the heated volume. This condition implies that there is negligible heat diffusion during the laser pulse so that the PA image essentially reveals the initial absorbed energy distribution in the sample. The second condition of "stress confinement" requires that time τ_s for stress to transit the heated volume should be longer than the laser pulse duration τ_p.

It can be shown that a nanosecond laser pulse impacting a biological tissue sample satisfies the conditions for PA imaging. Thermal diffusion length during the laser pulse is given by $\delta_T = 2(\tau_p D_T)^{1/2}$ where D_T is the thermal diffusivity of the

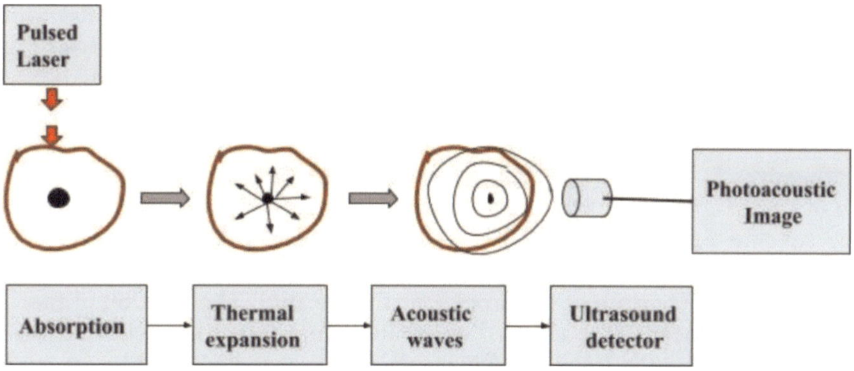

Figure 15.
Schematic illustration of the principle and processes involved in photoacoustic image formation.

sample. A typical value of $D_T = 1.4 \times 10^{-3} cm^2/s$ for most tissue [29] gives $\delta_T = 0.05\,\mu m$ for a laser pulse of $\tau_p = 5$ ns. Duration of thermal diffusion is given by $\tau_{th} = L^2/4D_T$ where L is the radius of the spherical shell of heat propagation in the sample. Thus for $L = 15\,\mu m$, we get $\tau_{th} = 0.4$ ms. If "v" is the velocity of sound in tissue, the time taken by the stress to transit a sample length L in the heated region is given by $\tau_s = L/v$. Assuming $v = 1.5\,mm/\mu s$ we get $\tau_s = 10$ ns. This simple calculation shows that we would achieve a spatial resolution of $15\,\mu m$ in the PA image if $\tau_s = 10$ ns and $\tau_{th} = 0.4$ ms. Thus we find that the process of heat generation by a 5 ns laser pulse would satisfy the two conditions for generating PA signals efficiently.

5.1 Photoacoustic microscopy and photoacoustic tomography

Photoacoustic tomography (PAT) and photoacoustic microscopy (PAM) are the two methods of PA imaging. In the PAT mode, an expanded laser beam illuminates the whole sample, and laser photons are absorbed at various points in the sample generating ultrasonic waves. PAT acquires depth-dependent information by time-of-flight measurements of the acoustic waves. An ultrasonic transducer placed outside the sample detects the PA signal, which is measured either by moving a single transducer around the sample or by using an array of transducers. PA image is obtained from the data set of PA signals by using appropriate reconstruction algorithms in the computer. In the case of PAM, the laser beam is focused into a tiny volume, and ultrasonic waves from this localized region are imaged by the detector. To obtain a 3D image, transverse sections at different depths in the sample are scanned in two dimensions. The axial resolution of the image can be as good as the optical resolution ($<1\,\mu m$) where the depth information is determined by the time of flight of ultrasonic waves.

5.2 Photoacoustic microscopy of zebra fish

Fluorescence microscopy is an effective tool in thin biological samples like single-celled organisms, but with slightly thicker samples, it becomes difficult to know where exactly the fluorescence originates. In a complex organism, like zebra fish, it is crucial to image deeper and deeper while the organism is kept alive. Fluorescent light emerging from the point of absorption suffers multiple scattering in the tissue on its way to the optical detector. This leads to loss of information on the origin and propagation path of the fluorescent light, giving rise to a blurred image and destruction of the spatial resolution. PA detection of optical absorption circumvents these limitations, because the sound waves travel through the diffuse biological media with much less distortion than light.

The experimental components used in PAM by Harrison et al. [30] are shown in **Figure 16a**. The laser beam, from a tunable source (L), is diverted down the 45° reflecting cone R, by the prism P. The curved surface of the cone directs the laser beam in a horizontal plane which is reflected downward by the corner polished surfaces inclined at 45° with the horizontal plane. The laser beam then travels downward along the longer sides of the ultrasound probe (T) to be focused about 10.5 mm below the bottom of the probe (see **Figure 16a**). The position of the ultrasound transducer (T) is adjusted vertically to match its focus with the laser focus by maximizing the PA signal from carbon fiber sample [30]. The combined ultrasound and laser light probe is mounted on a voice coil (VC) stage which is driven by a programmed motor controller to achieve oscillations at 10 Hz over a length of about 10 mm. The imaging system can be run in three different modes, (i) ultrasound mode, (ii) PA mode, and (iii) interlaced PA-ultrasound mode, and it can produce images of microvascular structures to depths of 2–3 mm in vivo as illustrated in **Figure 16b** for zebra fish.

In the recording of images of **Figure 16b**, the zebra fish was held on a rotating platform immersed in water. The position of the laser focus was fixed at a particular depth inside the body of the sample, and the platform was rotated through 360° to record the two-dimensional sections [31]. The location of fluorescent protein mCherry (in red) is clearly seen in the image of the zebra fish brain at the top, and transverse image slices of the zebra fish hindbrain are shown in the lower half of **Figure 16b**, where each slice is separated by a depth of 0.5 mm inside the tissue.

5.3 Photoacoustic imaging of prostate cancer

PA imaging is emerging as a new diagnosis technique with specificity, high resolution, and enough imaging depth for early detection of prostate cancer. Ex vivo multispectral PA imaging has been carried out to differentiate between malignant

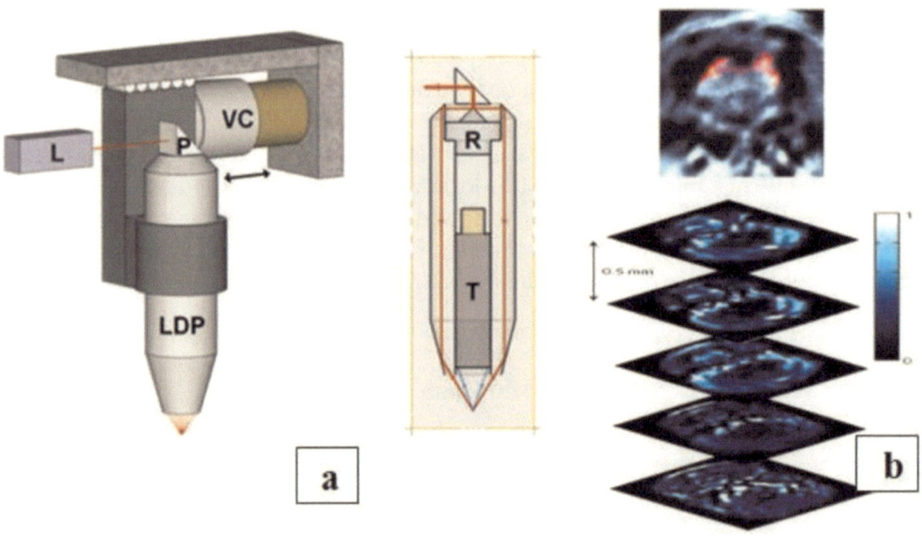

Figure 16.
(a) Experimental setup for combined photoacoustic (PA) and ultrasound (US) imaging (with permission from Ref. [30]). (b) In vivo image of a section of the zebra fish brain and five transverse image slices through the hindbrain (with permission from Ref. [31]).

prostate tissue, benign prostatic hyperplasia (BPH), and normal human prostate tissue. The preliminary results of investigations carried out by Dogra et al. [32] show that there was a significant difference in the mean PA intensity of dehydroxy hemoglobin (dHb) and lipid between malignant and normal prostate. There was also a significant difference in the mean intensity of dHb between malignant prostate and BPH. There was, however, no significant difference in HbO2, dHb, and lipid between normal prostate tissue and BPH. Laser radiation at 1064 nm and 1197 nm has been used to obtain PAT images, corresponding to optical absorption of hemoglobin and lipid, to determine the clustering prostate cancer tissue at each wavelength [33]. It was found that 1064 nm PAT in conjunction with ultrasound image is more effective in identifying prostate cancer biopsy targets than the PAT at 1197 nm.

6. Conclusion

This chapter starts with a brief history of photoacoustic effect and photoacoustic spectroscopy. A simple mathematical derivation for the generation of PA signal in gaseous and solid samples is followed by experimental methods. The design and construction of a variety of PA cells and detectors have been described along with their use in the investigation of gaseous, solid, and liquid samples. Some illustrative examples of trace detection of explosives and harmful chemicals have been discussed. A brief account of the principle and application of the emerging technique of PA imaging is discussed at the end of the chapter.

Acknowledgements

I am grateful to all the authors and scientists whose results have been cited to make the presentation meaningful. I am thankful to Dr. Punam Rai for taking care of my health, Sudheer for the help with the computer, and to my grandchildren, Leo and Mia, for their innocent inquiries during the course of my writing.

Author details

Surya Narayan Thakur
Banaras Hindu University, Varanasi, India

*Address all correspondence to: snthakur@gmail.com

References

[1] Bell AG. On the production and reproduction of sound by light. American Association for the Advancement of Science. 1880;**29**:115

[2] Kreuzer LB, Patel CKN. Nitric oxide air pollution: Detection by optoacoustic spectroscopy. Science. 1971;**173**:45

[3] Tyndall JG. Action of an intermittent beam of radiant heat upon gaseous matter. Royal Society of London. 1881; **31**:307

[4] Bell AG. Upon the production of sound by radiant energy. Philosophical Magazine. 1881;**11**:510

[5] Kasha M. Characterization of electronic transitions in complex molecules. Discussions of the Faraday Society. 1950;**9**:14

[6] Lewis GN, Kasha M. Phosphorescence and the triplet state. Journal of the American Chemical Society. 1944;**66**:2100

[7] Rosencwaig A, Gersho A. Theory of the photoacoustic effect with solids. Journal of Applied Physics. 1976;**47**:64

[8] Thakur SN. Photoacoustic spectroscopy of gases and powders. Bulletin of Laser and Spectroscopy Society of India. 1993;7:3

[9] Narayanan K, Thakur SN. Photoacoustic spectrum of iodine in the 490–680 nm (20402–14700 cm^{-1}) region. Applied Optics. 1992;**31**:4987

[10] Arnott WP, Moosmuller H, Rogers CF, Jin T, Bruch R. Photoacoustic spectrometer for measuring light absorption by aerosol: Instrument description. Atmospheric Environment. 1999;**33**:2845

[11] Lewis KA, Arnott WP, Moosmuller H, Wold CE. Strong spectral variation of biomass smoke light absorption and single scattering albedo observed with a novel dual-wavelength photoacoustic instrument. Journal of Geophysical Research. 2008;**113**:D16203

[12] Prasad RL, Prasad R, Bhar GC, Thakur SN. Photoacoustic spectra and modes of vibration of TNT and RDX at CO_2 laser wavelengths. Spectrochimica Acta. 2002;**A58**:3093

[13] Klenze R, Stumpe R, Kim JI. Laser-induced photoacoustic spectroscopy for the speciation of actinides in submicromol concentrations. In: Hess P, Pelzl J, editors. Photoacoustic and Photothermal Phenomena. Heidelberg: Springer-Verlag; 1988. p. 139

[14] Russo RE, Rojas D, Robouch PB, Silva RJ. Remote photoacoustic measurements in aqueous solutions using an optical fiber. The Review of Scientific Instruments. 1990;**61**:3729

[15] Kim JI. Actinide colloid generation in groundwater. Radiochimica Acta. 1991;**52/53**:71

[16] Kosterev AA, Bakhirikin YA, Curl RF, Tittle FK. Quartz-enhanced photoacoustic spectroscopy. Optics Letters. 2002;**27**:1902

[17] Weidmann D, Kosterev AA, Tittle FK, Ryan N, McDonald D. Application of widely electrically tunable diode laser to chemical gas sensing with quartzenhanced photoacoustic spectroscopy. Optics Letters. 2004;**29**: 1837

[18] Kosterev AA, Tittle FK. Ammonia detection using quartz-enhanced photoacoustic spectroscopy with a near-IR telecommunication diode laser. Applied Optics. 2004;**43**:6213

[19] Kosterev AA, Tittle FK. Chemical sensors based on quantum cascade

lasers. Journal of Quantum Electronics. 2002;**38**:582

[20] Van Neste CW, Senesac LR, Thundat T. Standoff photoacoustic spectroscopy. Applied Physics Letters. 2008;**92**:234102

[21] Altuzar V, Pacheco M, Tomás SA, Arriaga JL, Zelaya-Angel O, Sánchez-Sinencio F. Analysis of ethylene concentration in the Mexico City atmosphere by photoacoustic spectroscopy. Analytical Sciences (Special Issue) The Japan Society for Analytical Chemistry. 2001;**17**:s541

[22] Schmohl A, Miklos A, Hess P. Detection of ammonia by photoacoustic spectroscopy with semiconductor lasers. Applied Optics. 2002;**41**:1815

[23] Gustafsson LE, Leone AM, Persson MG, Wikuld NP, Moncada S. Endogenous nitric oxide is present in the exhaled air of rabbits, guinea pigs and humans. Biochemical and Biophysical Research Communications. 1991;**181**:852

[24] Lanz MJ, Leung DYM, McCormick DR, Harbeck R, Szefler SJ, White CW. Comparison of exhaled nitric oxide, serum eosinophilic cationic protein, and soluble interleukin-2 receptor in exacerbations of pediatric asthma. Pediatric Pulmonology. 1997;**24**:305

[25] Elia A, Lugara PM, Giancaspro C. Photoacoustic detection of nitric oxide by use of a quantum-cascade laser. Optics Letters. 2005;**30**:988

[26] Spagnolo V, Kosterev AA, Dong L, Lewicki R, Tittle FK. NO trace gas sensor based on quartz-enhanced photoacoustic spectroscopy and external cavity quantum cascade laser. Applied Physics. 2010;**B100**:125

[27] Harren FJM, Gotti G, Oomens J, Hekkert St L. Photoacoustic spectroscopy in trace gas monitoring. In: Meyers RA, editor. Encyclopedia of Analytical Chemistry. Chichester: John Wiley; 2000. pp. 2203-2226

[28] Prasad RL, Thakur SN, Bhar GC. CO_2 laser photoacoustic spectra and vibrational modes of heroin, morphine and narcotine. Pramana. 2002;**59**:487

[29] Duck FA. Physical Properties of Tissue: A Comprehensive Reference Book. London: Academic Press; 1990

[30] Harrison T, Ranasinghesagara JC, Lu H, Mathewson K, Walsh A, Zemp R. Combined photoacoustic and ultrasound biomicroscopy. Optics Express. 2009;**17**(412)

[31] Razansky D, Distel M, Vinegoni C, Ma R, Perrimon N, Koster RW, et al. Multispectral opto-acoustic tomography of deep-seated fluorescent proteins in vivo. Nature Photonics. 2009;**3**:412

[32] Dogra VS, Chinni BK, Valluru KS, Joseph JV, Ghazi A, Yao JL, et al. Multispectral photoacoustic imaging of prostate cancer: Preliminary ex-vivo results. Journal of Clinical Imaging Science. 2013;**3**:41

[33] Bungart BL, Lan L, Wang P, Li R, Koch MO, Cheng L, et al. Photoacoustic tomography of intact human prostates and vascular texture analysis identify prostate cancer biopsy targets. Photoacoustics. 2018;**11**:48

Chapter 3

A Balanced Slew-Rate High-Voltage Integrated Bipolar Pulse Generator for Medical Ultrasonic Imaging Applications

Chin Hsia

Abstract

This chapter describes the use of silicon-on-insulator (SOI) technology to develop balanced slew-rate pulse generators for medical ultrasound scanners, especially for multi-channel portable systems. Since ultrasonic transducers are usually composed of piezoelectric materials, most of which are capacitive, and the resonant frequency is usually in the order of tens of MHz, it is preferred to convert the high-frequency excited signals into high-voltage pulses to efficiently drive the transducers. In addition, the second harmonic leakage of the high-voltage pulse signal output by the pulse generator needs to be controlled such that the pulse generator can be applied to tissue harmonic imaging. Based on these considerations, the pulse generator architecture with balanced rising and falling edges proposed in this chapter is designed by synthesizing low-power, high-speed level shifters and a high-voltage H-bridge output stage to output high-voltage pulse signals with low harmonic distortion. The entire circuit integrates an 8-channel pulse generator, producing pulse signals >100 Vpp. The rise and fall times of the pulses are within 18.6 and 18.5 ns, respectively. The overall quiescent current is <60 uA (including the on-chip power supply). The output current drive capability is >2 A and the second harmonic distortion is as low as −40 dBc, indicating that the integrated pulse generator can be used in advanced, portable ultrasonic harmonic imaging systems.

Keywords: high-voltage pulse generator, ultrasound transmitter, medical imaging, harmonic distortion, low-power on-chip supply

1. Introduction

Compared with computed tomography (CT), X-ray and magnetic resonance imaging (MRI), medical ultrasound imaging is relatively fast, inexpensive, portable and radiation-free, and has become one of the most popular modalities in clinical examination. Currently, besides conventional B-mode and Doppler ultrasonic modality, advanced imaging formation method, such as photo-acoustic (PA) imaging, elasto-graphic imaging, as well as harmonic imaging is gradually becoming an important medical assessment tool for these echo-graphic systems. PA imaging offers unique capabilities in studying biological tissue based on optical absorption contrast [1]. This method converts a laser-induced ultrasonic signal into a measurable signal by means of a detector made of piezoelectric material. However, due to the sensitivity and bandwidth effects of the laser-excited ultrasonic signals to obtain high-quality and high-resolution PA images, they usually require broadband capabilities with frequencies above several hundred MHz, which is challenging in integrating PA imaging with conventional ultrasonic imaging modalities. Elastography is an imaging technology sensitive to tissue stiffness that was first described in the 1990s [2]. It has been further developed and improved to achieve a quantitative assessment of tissue stiffness. Elastography methods mainly utilize elastic changes in soft tissue caused by specific pathologies or physiological processes. However, currently the operator has to exert manual compression on the tissue with the ultrasound transducer to increase the imaging sensitivity. Manual compression works fairly well for superficial organs such as the breast and thyroid but is challenging for assessing elasticity in deeper located organs such as the liver, which practicality in medical instruments is not high. Harmonic imaging, however, based on the selective imaging of the harmonic frequency, especially using second harmonic signals for image formation [3]. Compared with PA imaging and elastographic imaging, second harmonic imaging can be easily implemented with conventional B-mode imaging and possesses higher lateral resolution and lower side lobes, and is, therefore, less sensitive to clutter and off-axis distortions [4].

However, since the echo signals detected by tissue harmonic imaging are generated by the nonlinear characteristics of the transmission medium according to the input fundamental wave signal, they are more susceptible to various harmonic sources. For example, the nonlinearity caused by the transmitted waveform, the bandwidth of the medical transducer, and the distorted echo signal can induce harmonic leakage and affect the overall performance of the imaging [5]. Several methods have been proposed in the literature to study the effects of harmonic leakage and to address sensitivity issues such as coded excitation, bandpass filter separation techniques [6] and pulse inversion [7]. These techniques have one thing in common, that is, they ultimately request Gaussian-type transmit signals with low harmonic content. Generating such signals typically requires a digital-to-analog converter (DAC) and a high-voltage linear power amplifier [8–10], which is not cost effective and power hungry, especially for compact devices applications such as portable ultrasound scanners. In order to reduce system power consumption, it is usually a good choice to use a pulse generator. However, as indicated in [11], the power of the second harmonic signal is at most 20 dB less than the power of the fundamental frequency signal. In fact, due to the bandwidth limitations of the ultrasonic transducer, this difference has become even larger [12]. Therefore, achieving the signal-to-noise ratio required for harmonic imaging requires the harmonic signal generated by the pulse generator itself as low as possible.

Generating low-power, low harmonic leakage, gated high-voltage pulse waves is not easy, and from a distortion perspective, the pulse generator must be carefully designed because the excitation device designed to drive the medical ultrasound transducer needs to be maintained a high voltage (>50–100 Vpp) output with faster switching capability than a few MHz. In most of the past, high voltage pulse generators implemented using discrete components for ultrasound imaging systems have demonstrated superior performance [13, 14]. Traditional off-the-shelf pulse generators typically operate below 10 MHz and are limited in size in multi-channel transmitter applications [15]. Recently, integrated pulse generators for multichannel, portable and compact ultrasonic scanners have been developed [16–20]. For these devices, if harmonic distortion is not considered, it can be easily developed using a high voltage CMOS compatible process. For example, bipolar pulse generators are less sensitive to even harmonic distortion and can produce lower harmonic pulses. However, if the transmitter is designed without regard to the slew rate of the rise and fall times and the positive and negative conduction times of the output voltage, a large second harmonic would be generated [21]. Some prior works employed multiple voltage-level pulse generators to reduce harmonic distortions [22–24], and these techniques usually require couples of voltage sources, thus increasing design complexity. In this chapter, we presented our design of a pulse generator architecture with a balanced slew-rate both in rising and falling edges to improve second harmonic distortion, and develop this pulse generator using CMOS-SOI technology for medical ultrasound scanners, especially for harmonic imaging applications.

This chapter is organized as follows. Section 2 considers the pulse excitation signals required in harmonic imaging systems. Section 3 describes the detailed design and circuit level implementation of the high voltage pulse generator. In Section 4, the experimental results of driving a medical grade piezoelectric transducer (PZT) with a designed pulse generator are given. The conclusion is given in Section 5.

2. Pulse excitation signals in harmonic imaging system

Since high-voltage excitation signals such as square waves or trapezoidal waveforms can efficiently drive ultrasonic transducers, harmonic components of such waveforms need to be considered in order to achieve low harmonic output for our applications. Given the nature of these ideal waveforms, Fourier analysis makes it easier to display their spectrum and explore the possibility of generating such a waveform. Basically, slew-rate limiting signals can reduce radiated emissions (electromagnetic interference and radio frequency interference) and harmonics of its fundamental portion. Therefore, we can consider a digital signals, $f(t)$, that are not ideal square waves, but may be approximated by trapezoidal waveforms with finite rise and fall times, t_r and t_f, as shown in **Figure 1**. This waveform employs characteristics of many digital signals, including clock pulses and PWM waveforms. In **Figure 1**, "A" stands for the amplitude of the signal, T_{period} is the signal's period, T_{on} and T_{off} set for the turn-on and turn-off time, respectively. The waveform can be expanded in a Fourier series as shown in Eq. (1) and the expansion coefficients are given by Eq. (2).

$$f(t) = C_0 + \sum_{n=1}^{\infty} C_n \cos(n\omega_0 t + \varnothing_{cn}) \tag{1}$$

$$C_n = \frac{1}{T}\int_{t_0}^{t_0+T} f(t)e^{-jn\omega_0 t}dt \tag{2}$$

$$\text{where } C_0 = \frac{AT_{on}}{T_{period}} - \frac{A}{2} \tag{3}$$

For $n \neq 0$ and if $t_r = t_f$, the magnitudes of the Fourier coefficients, C_n, are given by

$$C_n = 2\frac{AT_{on}}{T_{period}}\left|\frac{\sin\left(\frac{n\pi t_r}{T_{period}}\right)}{\frac{n\pi t_r}{T_{period}}}\right|\left|\frac{\sin\left(\frac{n\pi T_{on}}{T_{period}}\right)}{\frac{n\pi T_{on}}{T_{period}}}\right| \tag{4}$$

From (Eq. (4)), it can be observed that the waveform's harmonic energy at high frequencies is less than that of an ideal square wave if the rise and fall times are finite. Moreover, if the pulse has equal duty cycles for the turn-on and turn-off period, i.e.,

$$T_{On} = T_{off} = \frac{1}{2} T_{period} \tag{5}$$

$$\left| \frac{\sin\left(\frac{n\pi T_{on}}{T_{period}}\right)}{\frac{n\pi T_{on}}{T_{period}}} \right| = \left| \frac{\sin\left(\frac{1}{2} n\pi\right)}{\frac{n\pi}{2}} \right| \tag{6}$$

Equation (6) is equal to zero for even n, which stands for NO even harmonics when the duty cycle of the excitation pules is 50% (This is a reasonable assumption and can be easily achieved by generating such a signal.).

The value of t_r in Eq. (4) affects the higher order terms of C_n. While in the general case of $t_r = t_f$, the larger the t_r (t_f), the lower the harmonic terms at high frequency, which, in fact, has a low-pass filtering effect. When deriving the spectrum of this pulse train, the case of equal rise and fall times is often assumed. However, the difference between the rising edge and the falling edge causes excessive harmonics, and even if the duty ratio remains 50%, even harmonics still occurs due to such a nonlinear effect.

In order to model the effects of higher order harmonic terms at high frequencies due to non-linear rising and falling edges, a repetitive trapezoidal waveforms with slew-rate limited rise and fall times and a normalized −1.0 to +1.0 V peak-to-peak voltage swing can be modeled using Eq. (7). It is generally happened when the switching FET has a non-linear on-resistance, causing the output waveform an unsymmetrical rise and fall times, which is a phenomenon that often occurs with pulse generators. To simulate the effects of different slew-rates, the rise and fall times can be readjusted by changing the boundary conditions of each section in Eq. (7). The simulated bipolar pulse signals with unbalanced slew-rates are shown in **Figure 2** [21]. The pulses in **Figure 2** assume uniform amplitude envelopes. The bandwidth of such waveforms is inversely proportional to the time period, T_{period}, in **Figure 2**, and for simplified illustrations of slew-rate effects, the center frequency of the pulse signal is fixed to 2 MHz.

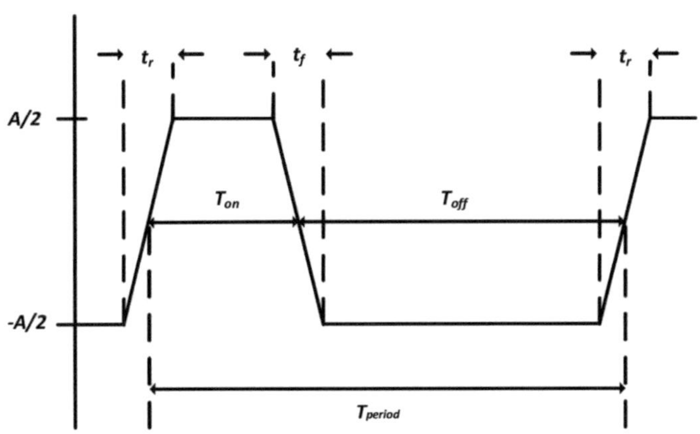

Figure 1.
A typical trapezoidal waveform or square waveform with the limited rise/fall time.

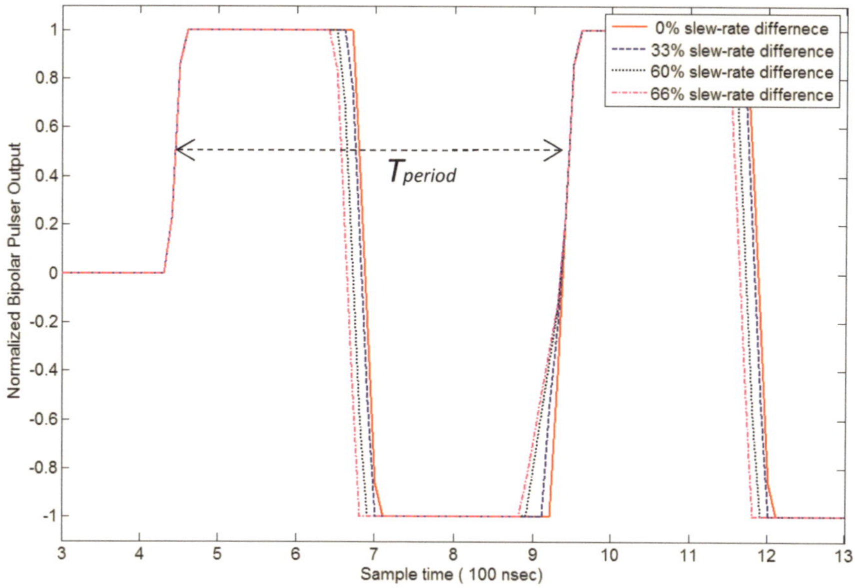

Figure 2.
Modeling of a bipolar pulse waveform with different slew-rates (rise/fall times) [21].

$$f_{trapezoid}(x) = \begin{cases} -4 - \dfrac{4x}{\pi} & \text{if } -\pi \le x < -3\pi/4 \\ -1 & \text{if } -3\pi/4 \le x < -\pi/4 \\ \dfrac{4x}{\pi} & \text{if } -\pi/4 \le x < +\pi/4 \\ +1 & \text{if } +\pi/4 \le x < +3\pi/4 \\ 4 - \dfrac{4x}{\pi} & \text{if } +3\pi/4 \le x < +\pi \end{cases} \tag{7}$$

In **Figure 2**, the slew-rate difference in percentage (SRDP) of the Rising_time and Falling_time can be defined in Eq. (8) as an important factor in evaluating signal harmonic damage. In Eq. (8), the Rising_time and Falling_time are defined as the pulse response to rise/fall from 10/90 to 90%/10% of its final values.

$$\text{SRDP} = \frac{|\text{Rising time} - \text{Falling time}|}{\text{Falling time}} \times 100\%. \tag{8}$$

Fourier analysis of this trapezoidal waveform provides harmonic magnitudes similar to that for an unsymmetrical square wave, except that there is an additional term that limits the high frequency content. **Figure 3** shows the spectrum of the excitation signals with equal on_time and off_time but different slew-rates, as shown in **Figure 2**, after mixing with the impulse function of the ultrasound transducer [21]. A unit-gain ultra-wideband transducer is assumed here without loss of generality to investigate the frequency components of the transmit signals. The spectrum shown in **Figure 3** presents that the second harmonic portion is damaged by signals with unequal slew-rates. For instance, the desired SRDP should be <33% in order to keep the second harmonics 40 dB lower than the fundamental tones.

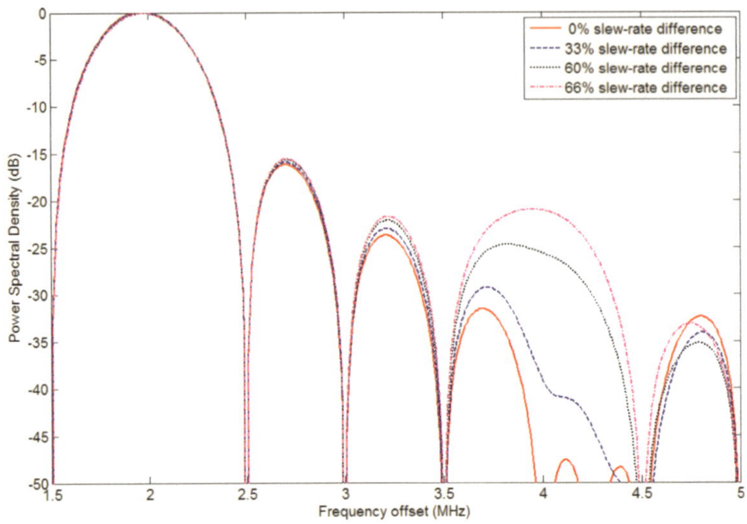

Figure 3.
The frequency spectra of the bipolar pulse waveforms with the different slew-rates (rise/fall time) [21].

Therefore, in order to maintain an excitation signal having a low second harmonic content, it is essential to control the pulse generator's slew-rate during the rising and falling edges.

3. High-voltage pulse generator for medical ultrasound

A diagnostic medical ultrasound image is acquired by transmitting acoustic waves and receiving echoes that are reflected from cell boundaries [20]. **Figure 4** shows a typical ultrasound analog front-end diagram comprises an HV pulse generator, an HV multiplexer (HV MUX), a T/R switch, a low noise amplifier (LNA), a variable-gain amplifier (VGA) and an analog-to-digital converter (ADC) [24]. The digital signal processor generates a sequence of low-voltage transmit signals with different delays for the pulse generator. The pulse generator drives the transducer at a high voltage level that, after exciting the array of ultrasonic probes, transmits a focused beam to the subject. Since the tissue of each part of the human body has discontinuous impedances to sound waves, the frequency and speed at which the transducer receives reflected sound energy from the body will result in distinguishable conditions. The T/R switch is used to reduce the interference at the receiving

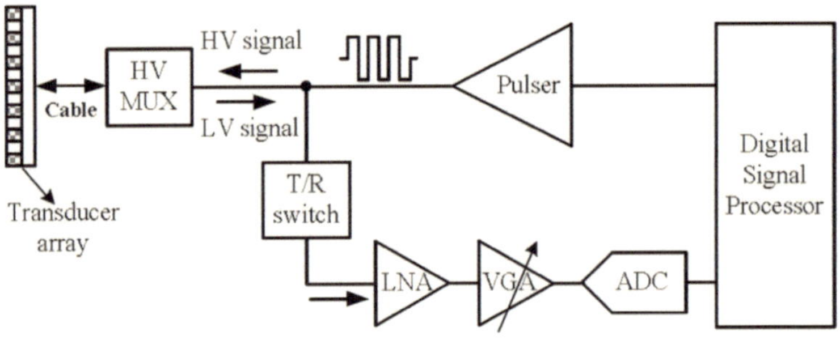

Figure 4.
A typical ultrasound system block diagram.

end of the LNA from the high voltage transmitter. When the pulse generator sends a high voltage transmitting pulse, the LNA is isolated from the pulse generator and the transducer by turning off the T/R switch, and when these echo signal is sent to the LNA via the T/R switch, the system is in receive mode, the pulse generator is disabled and the T/R switch is turned on, and the LNA and the follow-up VGA amplify the received signal and send the signal to the ADC. The digital signal processor generates information such as images or Doppler sound waves based on the output data of the ADC. In order to increase the strength of the echo signal, typically, multichannel transceivers are indispensable in modern ultrasound systems, which inevitably increase system complexity and power budget. The generation of multiple high voltage pulses, however, has been historically a challenge in the ultrasonic imaging system, particularly for today's portable solution. In the following section, we focused on the design and implementation of an integrated high-voltage pulse generator for multi-channel ultrasonic systems.

3.1 Single-chip high-voltage pulser architecture

The schematic of the integrated high-voltage pulse generator is shown in **Figure 5**. It mainly consists of three stages, including the input stage, on-chip floating power supplies, and H-bridge power driver. Several techniques are employed in this high-voltage pulse generator design to achieve low static power and low harmonics at the output. These techniques are described in the next section.

3.1.1 Input stage

The input stage, as shown in **Figure 6**, consists of signal conditioners (buffers), a dead time generator, and a return-to-zero controller. The input stage deals with the control of the pulse generator and produces HSDP, LSDN, RTZP, RTZN four signals to switch on/off the following H-bridge power driver. The two input control signals, IN1 and IN2, are generated separately from the outside signal generator or a DAC with synchronized phases. One of the control signals is a half cycle delay from the other to drive the push-pull stage at the output. The advantage of this architecture is that by adjusting the duty cycle of the input signals of IN1 and IN2, respectively, the error between on_time and off_time can be reduced, as shown in Eq. (6).

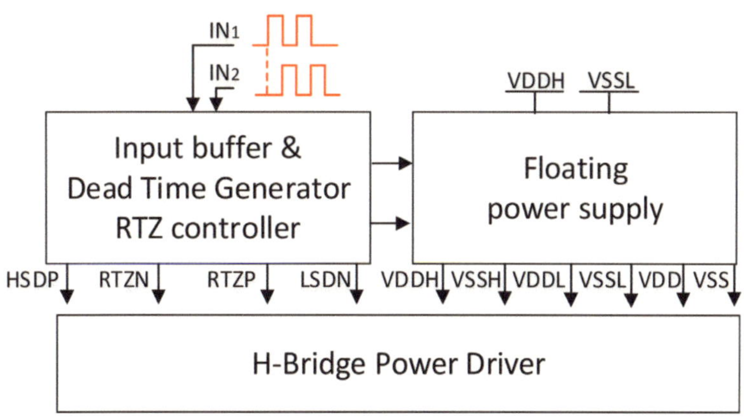

Figure 5.
Proposed architecture of the integrated bipolar pulse generator.

3.1.2 H-bridge power driver

The H-bridge power driver, as shown in **Figure 7**, composed of level shifters, pre-drivers, two high-voltage diodes (D1/D2), and four power transistors (MP1/2 and MN1/2). The low voltage and high voltage levels of the input control signal are 0 and 5 V, respectively. When the voltage level of IN1 is high, MN1 is turned off and MP1 is turned on. MP1 drives the output to a high voltage, VDDH. When the voltage level of IN2 is high, power transistor MN1 pulls the output down to a low voltage level VSSL. The voltage levels of IN1 and IN2 are low during steady state and the output returns to zero voltage level via MP2 or MN2. To exclude shoot-through current between MP1 and MN1, a dead-time control circuit is used as shown in **Figure 6**, in which, the outputs of the dead time generator are not simultaneously "high." Level-Shifter #1 and Level-Shifter #2 then boost these two output signals to the voltage levels of VDDL to VSSL and VSSH to VDDH, respectively. Due to the large capacitive load at the gates of the final stage power inverters MP1 and MN1, two pre-drivers are used to improve the slew rate. In addition, to avoid ringing of the output in a steady state, a fast return to zero signal is required. The return-to-zero circuit includes a return-to-zero controller, Level-Shifter #3, delay circuit, two pre-drivers, two power transistors (MP2 and MN2), and two high-voltage diodes (D1 and D2) to prevent false conduction. To avoid affecting the reception of echo signals when the ultrasound system is in receive mode, the output of the pulse generator must quickly return from the voltage level of VDDH or VSSL to zero voltage level. As shown in **Figure 6**, the return-to-zero controller can be composed of a simple XOR gate and an inverter.

Generating a pulse signal with low second harmonic contents is not easy, requiring balanced rising and falling times at all operating frequencies and very accurate duty cycles. In order to control the balanced slew-rate of the output pulses,

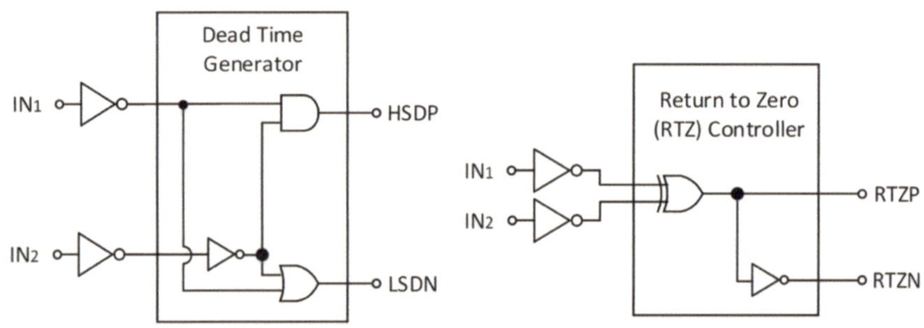

Figure 6.
Input stage of the bipolar pulse generator.

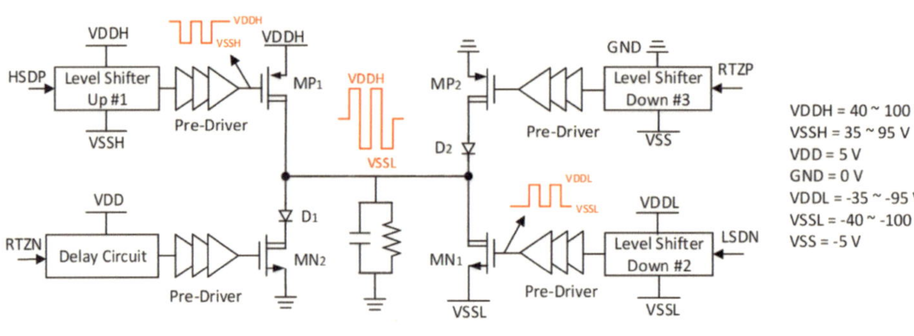

Figure 7.
The H-bridge power driver.

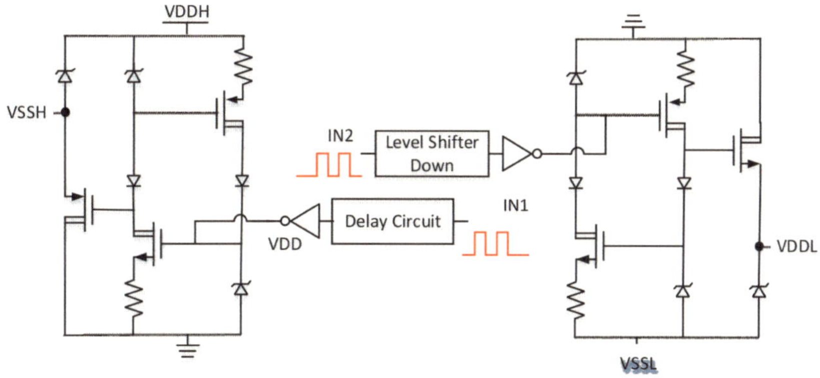

Figure 8.
Floating power supply schematic and its controller.

the rise/fall time of the level-shifters and output power inverters were carefully investigated. Sections 3.1.4 and 3.1.5 will describe in detail the design considerations of those key components.

3.1.3 Low-power floating power supply design

Since the output stage uses a push-pull architecture, a pair of high-voltage PMOS and NMOS devices are employed as the output stage. Design of their driver stages can also use push-pull architectures to efficiently drive the output stage. Since the gate voltage of high-voltage PMOS/NMOS devices is limited by the maximum operating voltage of Vgs, the output voltage swing of the driver has to be within 5 V. Therefore, each driver requires a set of 5 V rail-to-rail power supplies to provide a source/sink current to the output. In order to simplify the power supplies, a floating voltage source architecture is employed. **Figure 8** shows two floating voltage sources generated by the VDDH and VSSL to ground, respectively. A pair of high-voltage PMOS/NMOS devices in series with two sets of Zener and high-voltage diodes form a voltage divider loop and outputs two voltages respective to the voltage drop of the Zener diode. Another two high-voltage PMOS/NMOS in parallel with the divider loop is functioned as the source follower to provide stable output voltages. For instance, the output voltage VDDL and VSSH, are generated by VSSL and VDDH as shown in Eqs. (9) and (10), respectively.

$$\mathrm{VSSH} = \mathrm{VDDH} - \mathrm{V}_{\mathrm{Zener}} \tag{9}$$

$$\mathrm{VDDL} = \mathrm{VSSL} + \mathrm{V}_{\mathrm{Zener}} \tag{10}$$

One of the advantages of this architecture is its fast settling and stable output even though the output voltage is not accurately regulated. However, it dissipates power once the VDDH or VSSL is applied and degrade the efficiency of the overall system. Therefore, a switch generated by the input signals are employed. The floating power supply can only work following the input signal's commands. When there is no input signal for a period of time, the floating power supply can be turned off to reduce the power dissipation.

3.1.4 Capacitive-coupled floating level-shifter design

Level shifters are used in applications that require interfaces between different voltage domains. There are two types of level shifters: full-swing and floating, which can be distinguished by whether the voltage domain shares a common ground potential. **Figure 9** shows the schematic of a conventional latch-based full-swing level shifter, which is to boost the input signal from a "Vlow" level to a "Vhigh" level. The inverter chain (M1–M12) is used to reconstruct the rail-to-rail digital signal from the off-chip input control signal. The cross-couple pair (M15 and M16) can latch the "high" digital signal level. When the input signal switches to a "low" level, M16 and M17 are turned off, and M15 and M18 are turned on. The output voltage is 0 V. However, as the input signal becomes to a "high" level, M16 and 17 are switched on, and M15 and M18 are switched off. The output voltage becomes to Vhigh.

The full-swing level shifter shown in **Figure 9** is not appropriate for PMOS/NMOS gate driver design since the gate driver needs a floating rail to switch on/off the last stage of the PMOS/NMOS devices. Floating level shifters, however, can shift the potential of control signals from circuits of a low voltage power rail to the potential with floating and ground rails, and therefore, the floating level shifters are often used in the gate drivers to drive output stages. **Figure 10** shows the designed capacitive-coupled level shifter architecture employed for the H-bridge power driver.

Figure 10 shows the schematic of the floating level shifter for level shifter #1 to shift the signal voltage levels of 0 V and VDD to VSSH and VDDH, respectively. The level shifter consists of a pair of inverters (Mi1–Mi4), two coupling capacitors (C1 and C2), a latch (M19–M22), an output inverter (M23–M24), and a dummy inverter Mdum1 and Mdum2, which is to keep the same output impedance seen by inverters of M21 and M22. The pair of inverters is operated under 0 and 5 V power supplies, while the latch, the dummy inverter, and the output inverter are supplied with VSSH and VDDH power source. The pair of inverters and the latch are isolated by these two coupling capacitors, which couple through the control signals to the output stages. **Figure 11** shows the simplified model of the latched stage with the capacitive coupling between the input and the leveled outputs, in which, G_m represents the sum of the trans-conductance of M19 and M20 (same as M21 and M22). C_L is the input capacitance of M23 and M24. R_L represents the output node impedance

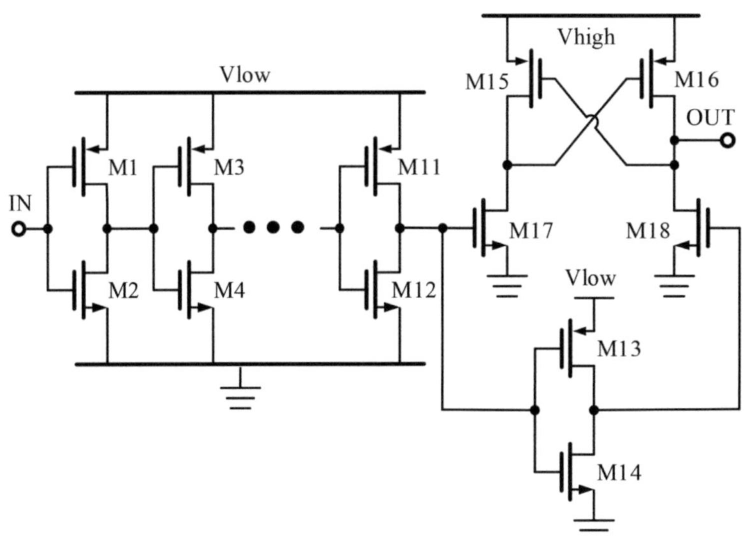

Figure 9.
Schematic of a latch-based level-shifter.

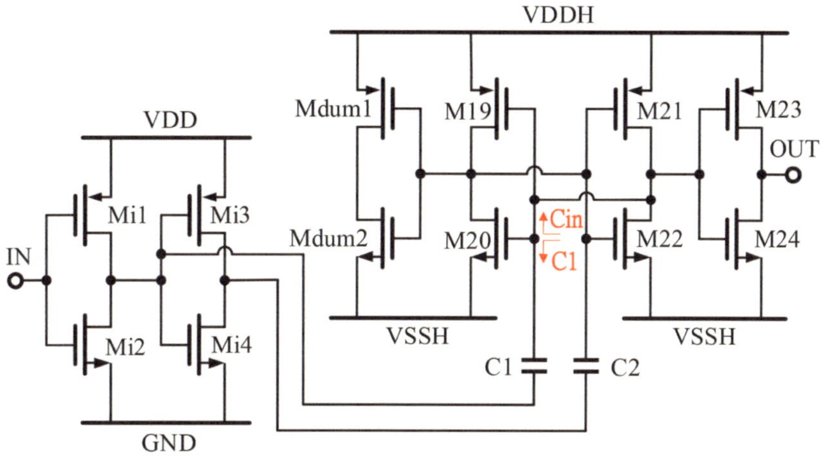

Figure 10.
Schematic of a latch-based capacitive-coupled floating level-shifter.

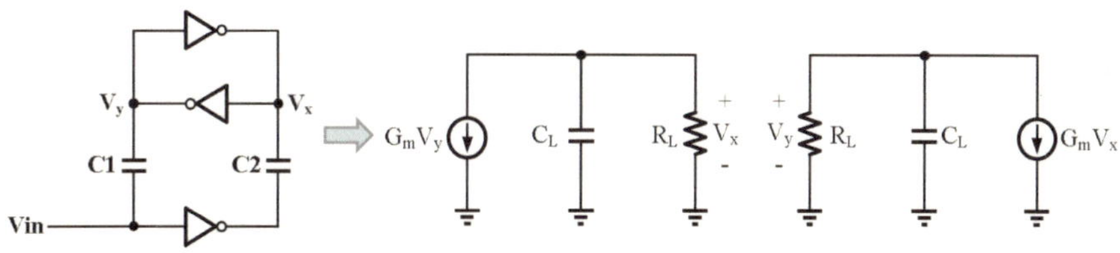

Figure 11.
Simplified schematic of the floating level shifter.

of M19 and M20. Following **Figure 11**, the dynamic behaviors of the latch output, V_x and V_y, can be written in Eq. (11) and Eq. (12).

$$G_m V_y = -C_L\left(\frac{dV_x}{dt}\right) - \left(\frac{V_x}{R_L}\right). \tag{11}$$

$$G_m V_x = -C_L\left(\frac{dV_y}{dt}\right) - \left(\frac{V_y}{R_L}\right). \tag{12}$$

By replacing R_L and C_L with $\tau = R_L C_L$, $A_v = G_m R_L$, and reorder the formula, Eqs. (13) and (14) represents the cross correlation between V_x and V_y.

$$\tau\left(\frac{dV_x}{dt}\right) + V_x = -A_v V_y. \tag{13}$$

$$\tau\left(\frac{dV_y}{dt}\right) + V_y = -A_v V_x. \tag{14}$$

From Eqs. (13) and (14), we can solve

$$\delta V = \delta V_0 * e^{\frac{(A_v-1)t}{\tau}} \tag{15}$$

where δV is the voltage difference between the input and output of the Latch, i.e., V_x–V_y in our design, and δV_0 is the initial voltage difference at the beginning of the latch phase. The transition time of the latch can be solved in Eq. (15) as

$$T_{latch} \cong \frac{C_L}{G_m} * \ln\left(\frac{\delta V}{\delta V_0}\right) \tag{16}$$

The slew-rate of the level-shifter can then be defined as in (Eq. (17))

$$SR^{+/-} = \frac{\delta V}{T_{latch}} \tag{17}$$

Since the latched time is reversed logarithmic proportional to $\delta V_0, T_{latch}$ will be too large to affect the desired slew-rate if δV_0 is a small value. From **Figure 10**, we can find $\delta V_0 = \frac{C_1}{C_1+C_{in}}$VDD; the capacitance ratio between the coupling capacitance and the input capacitance of the latches affects the latched time. The value of C_1 can be, therefore, designed comparable to that of C_{in} to avoid a small δV_0. In addition, in order to have a balanced slew-rate of the level-shifters, the coupling capacitance ratio of between C_1 and C_2 can be another important factor after the latches have been designed. **Figure 12** shows the simulated slew-rate of the level-shifter as the value of the coupling capacitance is varied accordingly, which the optimal design can be found by choosing the corresponding coupling capacitance with SR^+ equal to SR^-. Since the coupling capacitor has to withstand a large voltage drop between VDDH and VDD, on-chip MOM capacitors were employed in series to increase the withstand voltage to 100 V for such applications. **Figure 13** shows the schematic diagram and layout of the on-chip MOM capacitor.

3.1.5 Final stage power inverter design

The external slew rate of the pulse generator is also affected by the final stage design in addition to the balanced slew rate design of the level shifter. In order to coordinate the final slew-rate, the size of one power transistor, MN1 shown in **Figure 7**, is fixed, and the size of the other transistor, MP1, is swept accordingly. The output package effect is considered when sweeping the size of the transistor.

The package type used for the pulse generator is QFN-88L. The corresponding parasitic components are obtained using the ANSYS Q3D extractor. The outline of the QFN-88L package is shown in **Figure 14**. **Figure 15** shows the slew-rate of the rising and falling edges of the final stage as the device size of MP1 changes in the

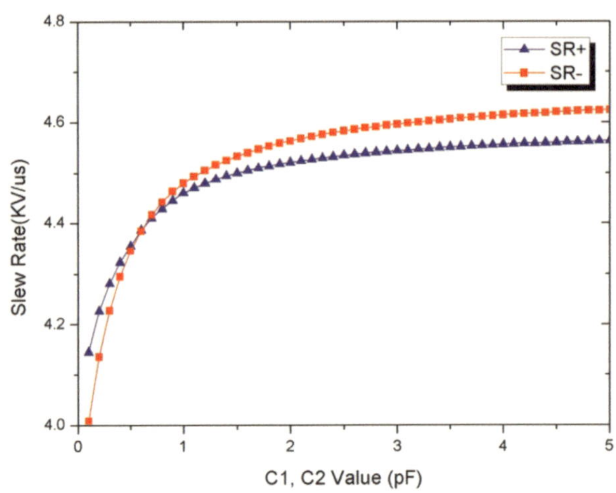

Figure 12.
Simulated slew-rate of rising and falling edge of the level-shifter.

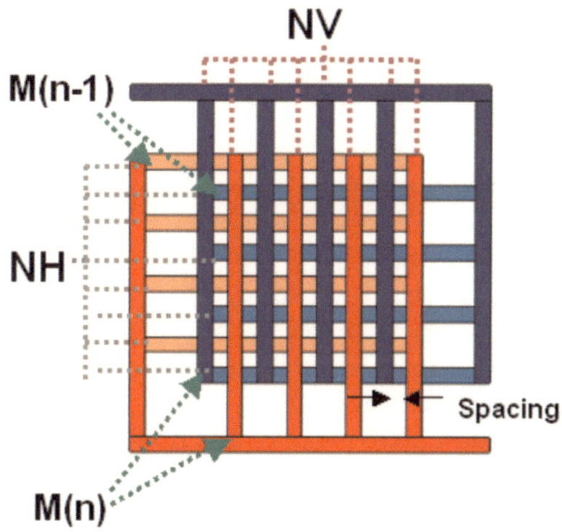

Figure 13.
Schematic of the on-chip MOM capacitor (M presents the layer of the overlapping fingers).

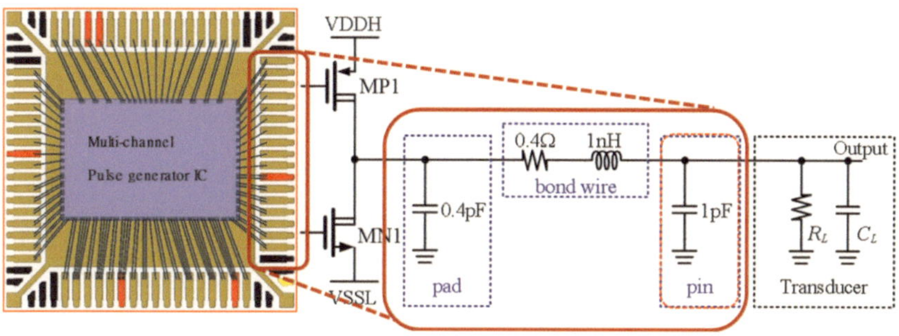

Figure 14.
Package/chip joint design with the parasitic parameter extractions.

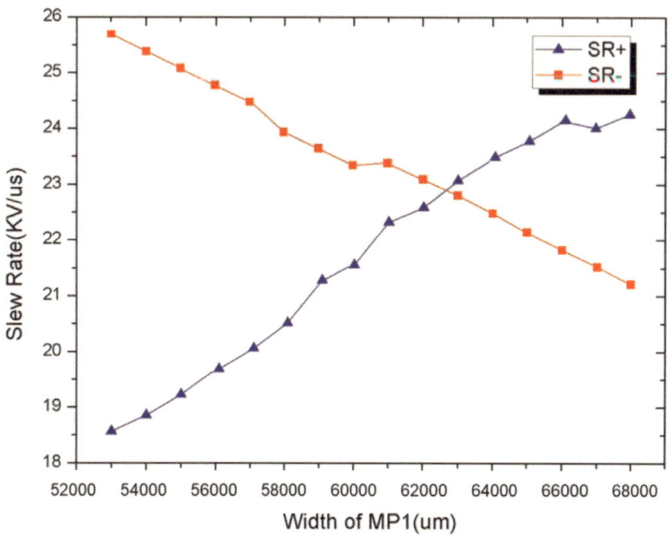

Figure 15.
Simulated slew-rate of rising and falling edge of the final stage MOSFETs.

simulated models. The optimum design width for this application is ~12,700 μm so that the final stage power MOSFETs operate at the same slew rate.

4. Experimental results

The proposed high-voltage pulse generator was fabricated in an 0.5 μm SOI-CMOS technology, which allows mixing different structures such as CMOS for digital circuits and high-voltage MOS structures for power and high-voltage applications on the same wafer with buried isolation layer [25]. **Figure 16** shows a cross-section of the SOI-CMOS process, and the high-voltage MOS transistors are fully compatible with the existing CMOS process. The final stage power transistors used

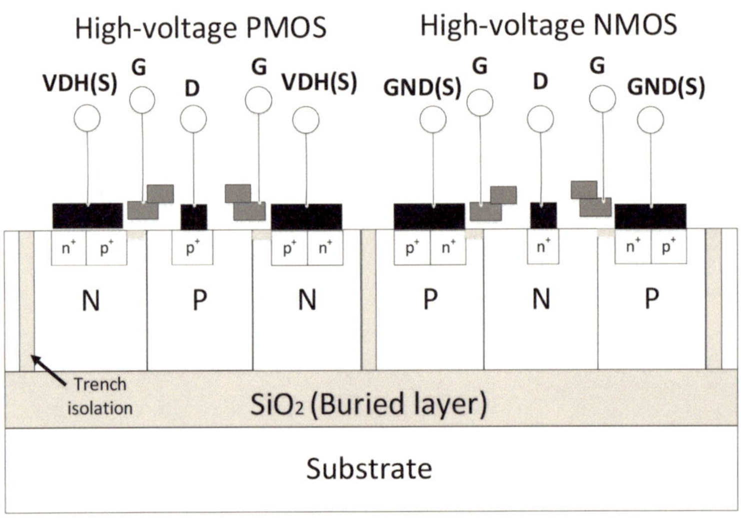

Figure 16.
The cross-section view of high-voltage devices in a SOI-CMOS process.

Figure 17.
The photo of the integrated 8-channel pulse generator.

in the design are 150-V N- and P-channel drain-extension field-effect transistors (FETs). The chip micrograph of the designed 8-channel pulse generator is shown in **Figure 17**, and it measures 8000 × 3510 μm area. The performance verification of the high voltage pulse generator was carried out by electrical and acoustic field measurement, which will be introduced in the next section.

4.1 Electrical performance verifications

The performance of the high voltage pulse generator is measured with a phased array probe using a 3.5 MHz gated input signal. The probe under test was Broadsound BS7L3, which has a fractional bandwidth of over 60%. **Figure 18** shows the measured bipolar voltage waveform and its spectrum diagram is shown in **Figure 19**. The second harmonic distortion is down to −40 dBc. The output voltage

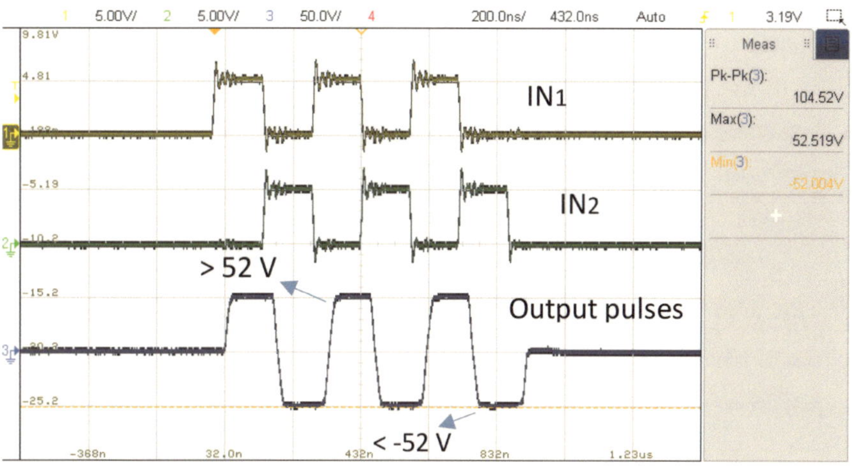

Figure 18.
Measured input signals and output waveform of the proposed pulse generator under a 1-K ohm resistance in parallel with a 220-pF capacitance load.

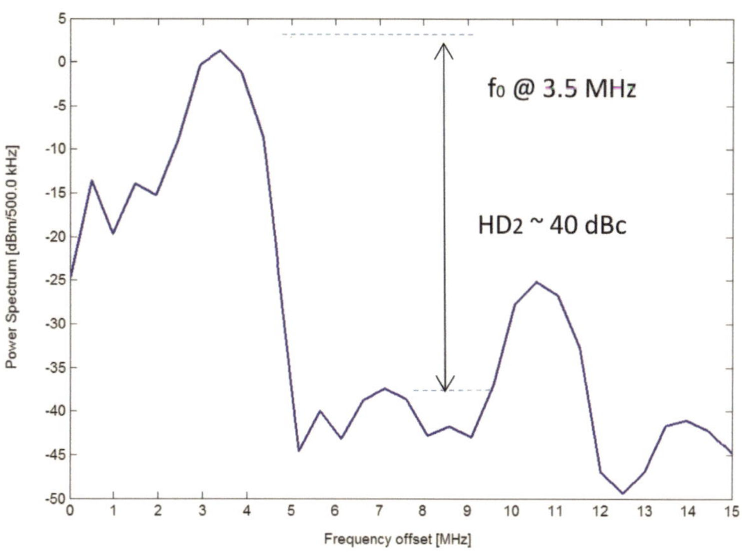

Figure 19.
*Spectrum analysis of the measured transmit waveform in **Figure 18**.*

	This work	LM96550	MAX14808	MD2130
Voltage level	3	3	5	Continuous
Technology	0.5 μm CMOS-SOI	N/A	N/A	CMOS (discrete)
Output voltage (V)	0 to ±70	0 to ±50	0 to ±100	0 to +125
Operating frequency (MHz)	up to 20	up to 15	up to 10	up to 15
Peak current (A)	> 2	2	2	3
Rising time (nsec)	18.6	18	21	-
Falling time (nsec)	18.5	18	21	-
Slew-rate (V/μsec)	4.4 K	2.2 K	3.8 K	-
Output load	1 kΩ//220 pF	100 Ω//330 pF	1 kΩ//240 pF	1 kΩ//220 pF
HD2 (dBc)	-40	-40	-43	-46
Standby power (mW)/channel	6	32	17	152

Table 1.
Measured performance summary and comparison.

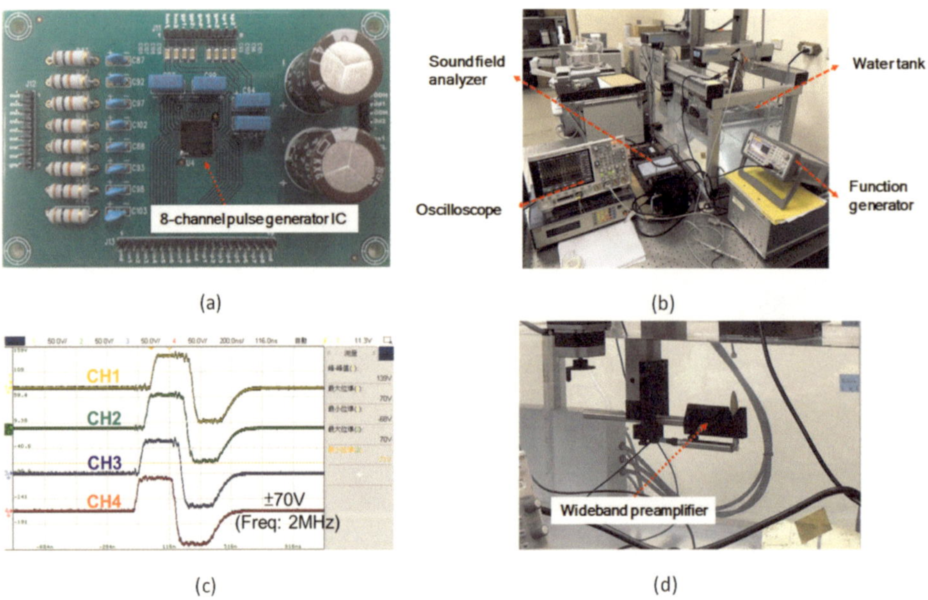

Figure 20.
The 8-channel pulse generator in a single package and its assembly (a), ONDA acoustic field calibration system (b), multiple-channel delayed high-voltage pulse output (c), and the wideband receiver of the sound field measurement (d).

can reach 100 Vpp with rising and falling times of 18.6 and 18.5 nsec, respectively. The static DC current provided to the pulse generator is about 60 uA per channel, which dissipates roughly 6 mW DC power. **Table 1** summarizes a performance comparison with several published works.

4.2 Acoustic field measurement results

The integrated 8-channel pulse generator test board for verifying transmit beamforming was assembled. The experiment was performed using an ONDA acoustic field calibration system. **Figure 20a** shows a picture of the designed 8-channel transmitter test board, and **Figure 20b** shows the picture of the measurement setup, mainly including a water tank to model the underwater environment and a sound filed analyzer to synthesize the beamforming results. After generating the input beamforming signals of different delay times by FPGA encoding, transducers are excited by these eight sets of high-voltage pulses. A wide dynamic range preamplifier is used to measure the sound field produced by the PZT probe after excitation by the pulse generator. **Figure 20c** presents the four-channel high-voltage output (±70 V) with certain preset delay. **Figure 20d** presents the wide-band receiver for the acoustic field measurements. **Figure 21** shows the measurements of transmit beamforming. The maximum 4 MPa was obtained at the focal

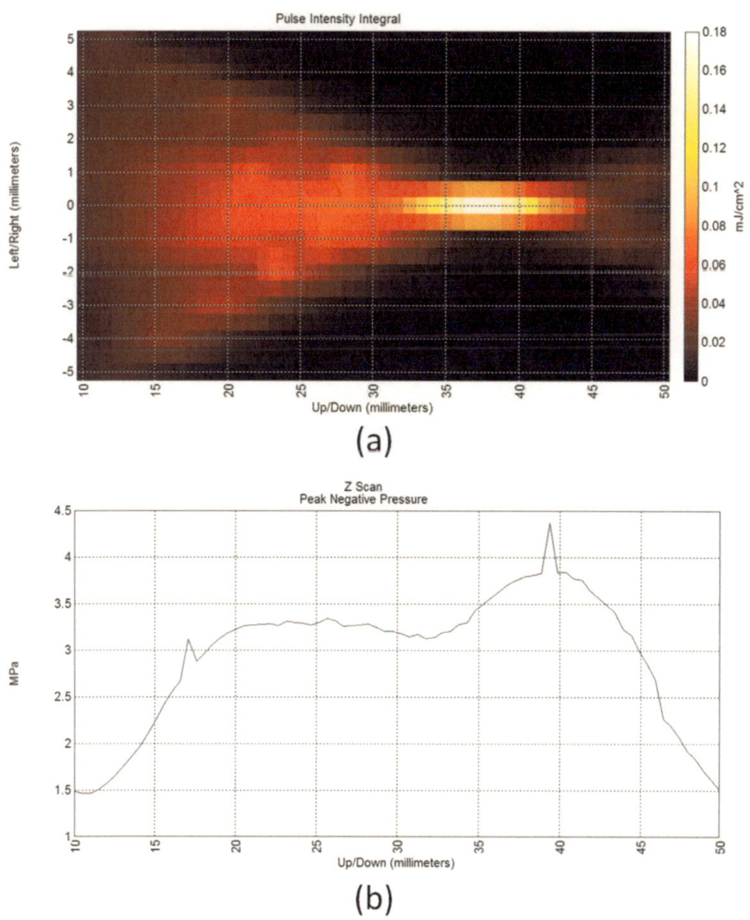

Figure 21.
Measured underwater acoustic field using the 8-channel pulse generator (a) and the maximum 4 MPa was obtained at the focal plane after beamforming (b).

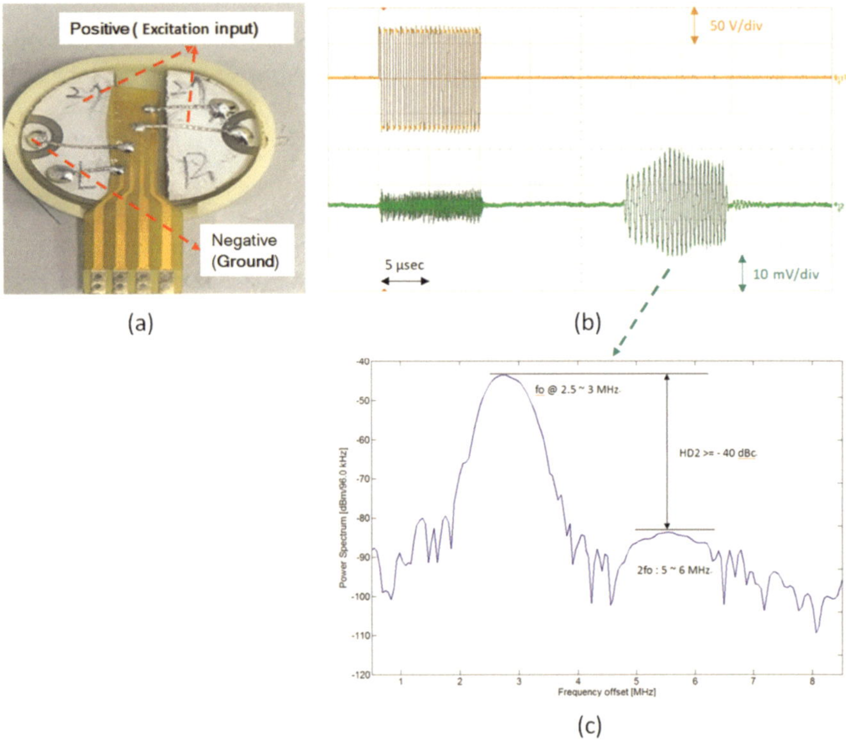

Figure 22.
The pulse-echo second harmonic leakage verification measurement setup with a patch transducer as the load of the pulse generator (a), high-voltage bipolar pulses as transmitting signals (orange) and the receiving signals (green) (b), and the spectrum analysis of the echo signals (c).

plane after beamforming. At a focal plane 3.5–4.0 cm away from the probe, the beam intensity is about 180 μJ/cm^2.

In order to verify the harmonic contents of the echo signal, another measurement setup was carried out using a single-channel patch transducer with the designed pulse generator. **Figure 22** shows the measured echo signals after exciting the patch transducer with the designed pulse generator. The existence of the received green signal under the high-voltage transmitted pulses is due to the leakage of the T/R switch. The second harmonic leakage at the receiver compared to the fundamental signal is <40 dB, making the pulse generator suitable for harmonic imaging applications.

5. Conclusions

Design and implementation of an integrated multiple channel high-voltage bipolar pulse generator for medical ultrasound transmitter applications were well addressed in this chapter. The proposed pulse generator can, assuming a capacitive ultrasound probe, generate output pulses >140 Vpp with rise and fall times of 18.6 and 18.5 ns, respectively. The resulting second harmonic distortion of the high-voltage pulse exceeds −40 dBc, meeting the requirements of ultrasonic transmitters in harmonic imaging applications. The performance of the integrated 8-channel ultrasonic pulses using the designed pulse generator was verified using the ONDA sound field calibration system. Experimental results show that the architecture can be used for advanced harmonic imaging ultrasound systems.

Author details

Chin Hsia
National Central University, Taoyuan, Taiwan, Republic of China

*Address all correspondence to: chinhsiancu2014@gmail.com

References

[1] Dong B, Sun C, Zhang HF. Optical detection of ultrasound in photoacoustic imaging. IEEE Transactions on Biomedical Engineering. 2017;**64**(1):4-15

[2] Sigrist RMS, Liau J, El Kaffas A, Chammas MC, Willmann JK. Ultrasound elastography: Review of techniques and clinical applications. Theranostics. 7 Mar 2017;**7**(5):1303-1329. DOI: 10.7150/thno.18650

[3] Averkiou MA. Tissue harmonic imaging. In: IEEE Ultrasonics Symposium; 2000

[4] Trucco A, Bertora F. Harmonic beamforming: Performance analysis and imaging results. IEEE Transactions on Instrumentation and Measurement. 2006;**55**(6):1965-1974

[5] Shen C-C, Li P-C. Harmonic leakage and image quality degradation in tissue harmonic imaging. IEEE Transactions on Ultrasonics, Ferroelectrics, and Frequency Control. 2001;**48**(3):728-736

[6] Song H-J, Song J, Chang J-H, Song T-K. Ultrasound harmonic imaging using nonlinear chirp for cardiac imaging. In: IEEE International Ultrasonics Symposium Proceedings; 2010

[7] Shen CC, Chou YH, Li PC. Pulse inversion techniques in ultrasonic nonlinear imaging. Journal of Ultrasound in Medicine. 2005;**13**(1):3-17

[8] 25A250A, 25 Watt CW, 10 kHz-250 MHz RF Power Amplifier, Amplifier Research

[9] ENI325LA, RF Power Amplifier, 250 kHz to 150 MHz, 25 W, 50 dB, Electronic Navigation Industries

[10] MD2130, High Speed Ultrasound Beamforming Source Driver, Supertex; 2011

[11] Song J, Kim S, Sohn HY, Song TK, Yoo YM. Coded excitation for ultrasound tissue harmonic imaging. Ultrasonics. 2010;**50**:613-619

[12] Cobbold RS. Foundations of Biomedical Ultrasound. Oxford, New York: Oxford University Press, Inc.; 2007

[13] Brown JA, Lockwood GR. A low-cost, high-performance pulse generator for ultrasound imaging. IEEE Transactions on Ultrasonics, Ferroelectrics, and Frequency Control. 2002;**49**(6):848-851

[14] Xu X, Yen JT, Shung KK. A low-cost bipolar pulse generator for high-frequency ultrasound applications. IEEE Transactions on Ultrasonics, Ferroelectrics, and Frequency Control. Feb 2007;**54**(2):443-447

[15] LM-96550, Ultrasound Transmit Pulser. Texas Instruments; 2013

[16] Haider B. Power drive circuits for diagnostic medical ultrasound. In: 2006 IEEE International Symposium on Power Semiconductor Devices and IC's; 2006

[17] Banuaji A, Cha HK. A15-V bidirectional ultrasound interface analog front-end IC for medical imaging using standard CMOS technology. IEEE Transactions on Circuits and Systems II. 2014;**61**(8):604-608

[18] Zhao D, Tan MT, Cha HK, Qu J, Mei Y, Yu H, et al. High-voltage pulser for ultrasound medical imaging applications. In: 13th International Symposium on Integrated Circuits; 2011. pp. 408-411

[19] Tan M, Chen C, Chen Z, Janjic J, Daeichin V, Chang Z-Y, et al. A front-end ASIC with high-voltage transmit

switching and receive digitization for 3-D forward-looking intravascular ultrasound imaging. IEEE Journal of Solid-State Circuits. Aug 2018;**53**(8): 2284-2297

[20] Hsia C, Huang Y, Lu C-W. Single-chip ultra-high slew-rate pulse generator for ultrasound scanner applications. In: 2013 IEEE International Ultrasonics Symposium (IUS); 2013. pp. 21-25

[21] Lu DF, Hsia C, Huang YC. Design of integrated high voltage pulse generator for medical ultrasound transmitters. IEICE Transactions on Communications. June 2019;**E102-B**(6): 1121-1127

[22] MAX14808, Octal Three-Level/ Quad Five-Level High-Voltage 2A Digital Pulsers with T/R Switch. Maxim Integrated Products Inc.; 2013

[23] HDL6V5540, Quad 5-level RTZ, 100V, 2A Ultrasound Pulser. Hitachi Ltd.; 2010

[24] Huang YC, Chen KT, Lu KY, Hsia C. 8-Channel High-Voltage Pulse Drive Circuit for Diagnostic Medical Ultrasound. SEMBA; 2019

[25] Qiao M, Zhang K, Zhou X, Zou J, Zhang B, Li Z. 250 V thin-layer SOI technology with field pLDMOS for high-voltage switching IC. IEEE Transactions on Electron Devices. 2015; **62**(6):1970-1976

Chapter 4

Determination of the Source Localization and the Beginning Time of the Acoustic Signal

Rostyslav Romanyshyn, Galyna Romanyshyn and Igor Romanyshyn

Abstract

For the problem of PAT, that is, determining the coordinates and the beginning time of an acoustic signal, it is necessary to carry out synchronized registration of acoustic signals of a source using a multichannel receiving system. Synchronously recorded signals are the signals with a delayed (long) front. A threshold method is proposed for determining the arrival time of noisy acoustic signals with a delayed front based on the evaluation of an adaptive threshold. An approach that allows to reduce the problem of determining the coordinates and the beginning time of an acoustic signal to solving a system of linear algebraic equations is proposed. Matrix A of the system of linear algebraic equations depends on the arrival times of synchronized registered signals (source coordinates). Therefore, when collecting data for a given geometry of the product and the location of the receivers, it is necessary to calculate areas, where matrix A is ill-conditioned. Areas of poor conditionality of matrix A should be excluded from the permissible areas of location of sources of acoustic signals. For these areas there will certainly be poor accuracy. The results of simulation and experimental testing of the developed PAT technologies are presented.

Keywords: acoustic signal, multichannel receiving system, source localization, signal processing, linear algebraic equations, coefficient of conditionality of matrix

1. Introduction

The key technologies of PAT are:

- Synchronized registration of acoustic signals of sources by multichannel receiving system
- Determination of the arrival times of the acoustic signals with the long (delayed) front to the receivers of the multichannel receiving system
- Calculation of the source localization and beginning time of the acoustic signal

- Calculation of the projection data—the propagation times of the acoustical signal (averaged slowness) along the propagation paths from the source to the receivers of the multichannel receiving system

- Tomographic reconstruction of the spatial distribution of the acoustic characteristics of the material

If necessary, iterative refinement of the source localization of acoustic signals and the spatial distribution of the acoustic characteristics of the medium is possible.

2. Synchronized recording of the acoustic signals

An important stage of the PAT is the synchronized recording of acoustic signals by a multichannel receiving system, which is necessary to determine the coordinates of the source and the beginning time of the signal.

Acoustic and electromagnetic synchronization of the recording of acoustic signals by a multichannel receiving system is used. In [1] a system for determination of localization of cracks from acoustic emission signals in which synchronization of recording is achieved by using an electromagnetic emission pulse is described.

The acoustic synchronization of the recording of acoustic signals consists in the following. When a signal arrives at the nearest detector of the multichannel receiving system, all other detectors start registering acoustic signals.

Synchronized recorded acoustic signals are the signals with delayed front. They consist of two sections. The first section is a noise segment that is associated with the time of propagation of an acoustic signal from a source to a receiver. This section is used to determine projections (travel times or slowness averaged over propagation paths). The second section is the noisy acoustic signal. This section is used for determination of the signal travel time, which is used to calculate of the source localization and beginning time of the signal.

3. The method for determination of the arrival time of an acoustic signal with a delayed front

The measured parameters by PAT are the arrival times of synchronized registered acoustic signals by a multichannel receiving system.

The techniques of determination of the signal arrival type are dependent on this signals nature, particular on the delayed front that is caused by the noise segment at its beginning.

In **Figure 1** an example of a particular time realization is shown, which is registered from one of the channels of an eight-channel acoustic signals recording system that were caused with a short blow with a scalpel on a steel plate.

To determine the arrival time of noisy signals with a delayed front, several methods have been developed [1–3].

A new method for determining the arrival time of a signal with a noisy front section was developed. The technique takes the attenuation of signals into account.

Let the implementation of the acoustic signal be a set of discrete samples:

$$s = \{s(n)|n = 1, \dots, N\}. \tag{1}$$

We choose the averaging filter window Δ and perform averaging of the signal:

$$z = \left\{ z(m) = \frac{1}{\Delta} \sum_{k=m-\frac{\Delta}{2}}^{m+\frac{\Delta}{2}} s(k) \middle| m = \frac{\Delta}{2} + 1, \ldots, N - \frac{\Delta}{2} \right\}. \tag{2}$$

Then we calculate the modulus of changes between consecutive samples of realization z:

$$d = \left\{ d(m) = |z(m+1) - z(m)| \middle| m = \frac{\Delta}{2} + 1, \ldots, N - \frac{\Delta}{2} - 1 \right\}. \tag{3}$$

After this, we choose the averaging filter window and perform averaging of the realization:

$$y = \left\{ y(m) = \frac{1}{\Delta_d} \sum_{k=m-\frac{\Delta_d}{2}}^{m+\frac{\Delta_d}{2}} d(k) \middle| m = \frac{\Delta}{2} + \frac{\Delta_d}{2} + 1, \ldots, N - \frac{\Delta}{2} - \frac{\Delta_d}{2} \right\}. \tag{4}$$

Further, we construct the line of a threshold that linearly decreases from p_m to 0 as the reading number m increases from $m_{\min} = \frac{\Delta}{2} + \frac{\Delta_d}{2} + 1$ to $m_{\max} = N - \frac{\Delta}{2} - \frac{\Delta_d}{2}$:

$$p = \left\{ p(m) = p_m \left(1 - \frac{m - \frac{\Delta}{2} - \frac{\Delta_d}{2} - 1}{N - \Delta - \Delta_d - 1} \right) \middle| m = \frac{\Delta}{2} + \frac{\Delta_d}{2} + 1, \ldots, N - \frac{\Delta}{2} - \frac{\Delta_d}{2} \right\}. \tag{5}$$

The signal arrival time is then taken to be the reading m_s for which:

$$p(m_s) = y(m_s). \tag{6}$$

The parameters Δ, Δ_d, p_m were selected based on experimental research.

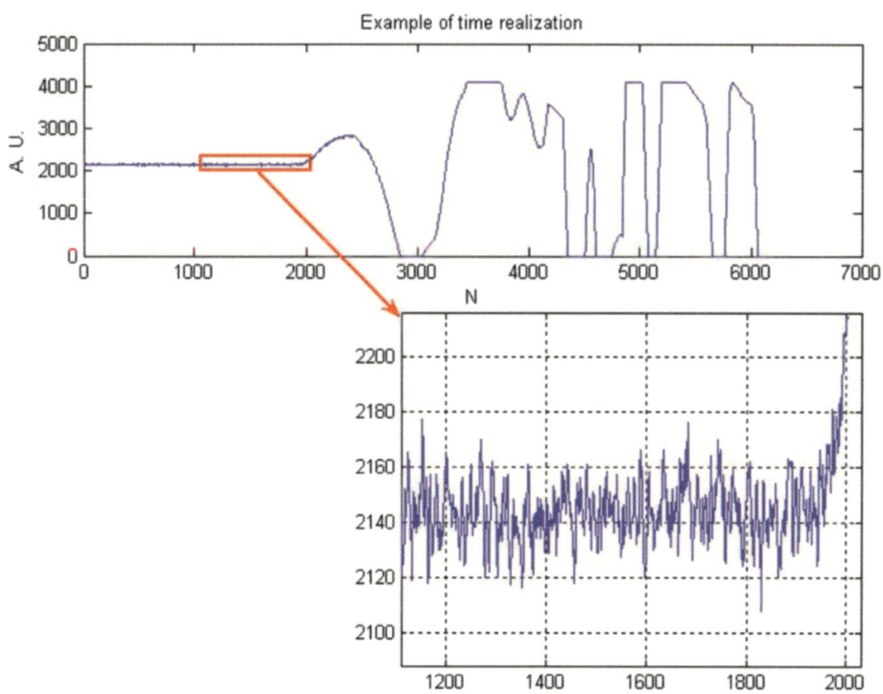

Figure 1.
An example of a temporal implementation of a registered acoustic signal.

When the arrival times of acoustic signals synchronized registered by the recording system have been determined, the problem is to calculate the source localization and beginning time of the acoustic signal.

4. Determination of the source localization and beginning time of an acoustical signal

There are a number of methods of localization of acoustic sources [4–6].

Let us consider a 2D case and assume that we use a three-channel recording system to determine the source localization and beginning time of an acoustic signal.

Let (x_0, y_0) be unknown coordinates of the source of acoustic signals and $(x_1, y_1), (x_2, y_2), (x_3, y_3)$ be the coordinates of the detectors. For the uniform distribution of the propagation velocity of acoustic signal, we have:

$$\begin{aligned} c(t_1 - t_0) &= \sqrt{(x_1 - x_0)^2 + (y_1 - y_0)^2} \\ c(t_2 - t_0) &= \sqrt{(x_2 - x_0)^2 + (y_2 - y_0)^2}, \\ c(t_3 - t_0) &= \sqrt{(x_3 - x_0)^2 + (y_3 - y_0)^2} \end{aligned} \tag{7}$$

where t_1, t_2, t_3 are the arrival times of the synchronized detected signals at the detectors and t_0 is the beginning time of the acoustic signal.

In practical terms, we can only determine the times of delay in the signal's arrival at one detector with respect to another $t_1 - t_2, t_1 - t_3$. Based on Eq. (7), we obtain the following system of two equations with two unknown coordinates of the source of acoustic signals:

$$\begin{aligned} c(t_1 - t_2) &= \sqrt{(x_1 - x_0)^2 + (y_1 - y_0)^2} - \sqrt{(x_2 - x_0)^2 + (y_2 - y_0)^2} \\ c(t_1 - t_3) &= \sqrt{(x_1 - x_0)^2 + (y_1 - y_0)^2} - \sqrt{(x_3 - x_0)^2 + (y_3 - y_0)^2} \end{aligned}. \tag{8}$$

From the geometrical viewpoint, solving Eqs. (7) or (8) reduces to finding the point of intersection of several circles or hyperbolae, respectively.

We present an approach that allows us to reduce the problem of determination of the coordinates of the source and the beginning time of an acoustic signal to solving of a system of linear algebraic equations.

The source coordinates (in 3D-case) and beginning time of acoustic signals are determined based on detecting of signal arrival times by four receivers:

$$\begin{aligned} (x_n - x_e)^2 + (y_n - y_e)^2 + (z_n - z_e)^2 &= c^2(t_n - t_e)^2, \\ (x_k - x_e)^2 + (y_k - y_e)^2 + (z_k - z_e)^2 &= c^2(t_k - t_e)^2, \\ (x_l - x_e)^2 + (y_l - y_e)^2 + (z_l - z_e)^2 &= c^2(t_l - t_e)^2, \\ (x_m - x_e)^2 + (y_m - y_e)^2 + (z_m - z_e)^2 &= c^2(t_m - t_e)^2 \end{aligned} \tag{9}$$

where $(x_i, y_i, z_i; t_i)$ are the coordinates of the receivers and the recorded signal arrival time $(i = n, k, l, m)$ and $(x_e, y_e, z_e; t_e)$ are the sought-for source coordinates and beginning time of the acoustic signal. The z_i coordinate is the same for all the receivers that are situated on the surface of the test article. Let n be the ordinal number of the receiver that was the first to register the signal. If the receivers are positioned equidistant on the sides of the article (which is taken to be a rectangle), we have $k = n + N/4$, $l = n + N/2$, and $k = n + 3/4N$, where N is the number of receivers that is a multiple of four.

After subtracting the first equation from the second equation, the first equation from the third equation, and the first equation from the fourth equation for a homogeneous medium (c = const), we arrive at the system of three linear algebraic equations:

$$As = b. \tag{10}$$

Here $A = \begin{pmatrix} 2(x_k - x_n) & 2(y_k - y_n) & 2c^2(t_n - t_k) \\ 2(x_l - x_n) & 2(y_l - y_n) & 2c^2(t_n - t_l) \\ 2(x_m - x_n) & 2(y_m - y_n) & 2c^2(t_n - t_m) \end{pmatrix}$; $s = (x_e, y_e, t_e,)^T$ is the sought-for vector, and $b = \begin{pmatrix} (x_k{}^2 - x_n{}^2) + (y_k{}^2 - y_n{}^2) + c^2(t_n{}^2 - t_k{}^2) \\ (x_l{}^2 - x_n{}^2) + (y_l{}^2 - y_n{}^2) + c^2(t_n{}^2 - t_l{}^2) \\ (x_m{}^2 - x_n{}^2) + (y_m{}^2 - y_n{}^2) + c^2(t_n{}^2 - t_m{}^2) \end{pmatrix}$. The coordinate z_e is determined from Eq. (9).

If the thickness of the article is considerably smaller than its length or width, we can take z_e to be half of the article thickness.

The matrix A depends on the arrival times of synchronously detected signals (the source coordinates). Therefore the allocation of the area where matrix is ill-conditioned should be determined when gathering data on the test article with the prescribed geometrical parameters and known receivers disposition (x_e, y_e, z_e).

The condition number of a matrix is given by [7]:

$$k = \frac{|\lambda_{\max}|}{|\lambda_{\min}|}, \tag{11}$$

where $|\lambda_{\max}|$ and $|\lambda_{\min}|$ are the maximal and minimal (by moduli) eigenvalues of A, respectively.

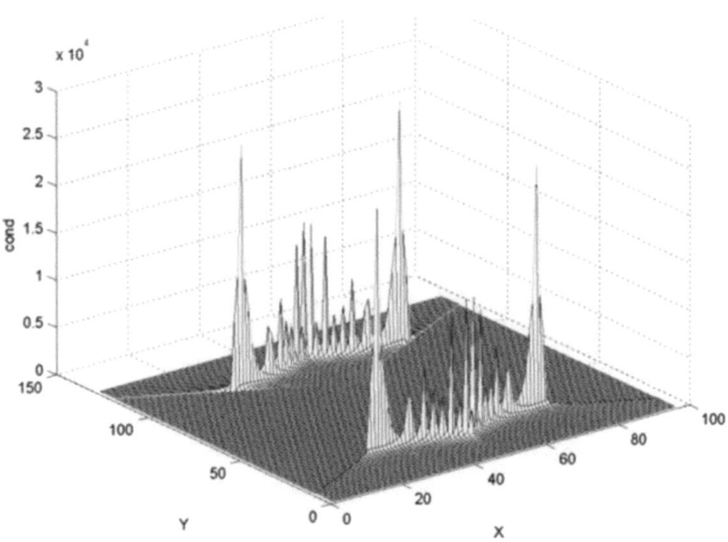

Figure 2.
Dependence of the condition number of the matrix on source coordinates.

Figure 2 provides an example of the dependence of the matrix condition number on the source coordinates (x_e, y_e) for a plane-parallel object with dimensions $96 \times 128 \times 8$ mm, with the receivers situated on the object surface at the centers of the sides. The domains where the matrix A is ill-conditioned must be excluded from admissible domains for the location of source of acoustic signals.

5. Description of the experimental setup and results of experimental testing

An experimental bench (an eight-channel acoustic emission tomograph [8]) was developed for experimental testing of the techniques of passive ultrasonic tomography. It is based on an eight-channel recording system that allows synchronized detection of acoustic signals with a 12-bit ADC with a sampling rate of 10–65 MHz. Acoustic and electromagnetic synchronizations for detection are provided in the tomograph. Experimental testing of the techniques was performed with a thick sheet St20 plate ($480 \times 640 \times 30$ mm) with a welded joint across the middle of the shorter side (at $x = 240$ mm, if the x axis is directed along the shorter side). Wideband DISM1 acoustic emission transducers [9] with a transmission bandwidth of 200–1200 kHz manufactured at the Bakul Institute of Superhard Materials of NAS of Ukraine were used for registration of acoustic signals. The transducers were attached to the butt ends of the sides at "points" with the coordinates (in mm) (160, 0); (320, 0); (480, 640/3); (480, $640 \times 2/3$); (320, 640); (160, 640); (0, 640/3); and (0, $640 \times 2/3$).

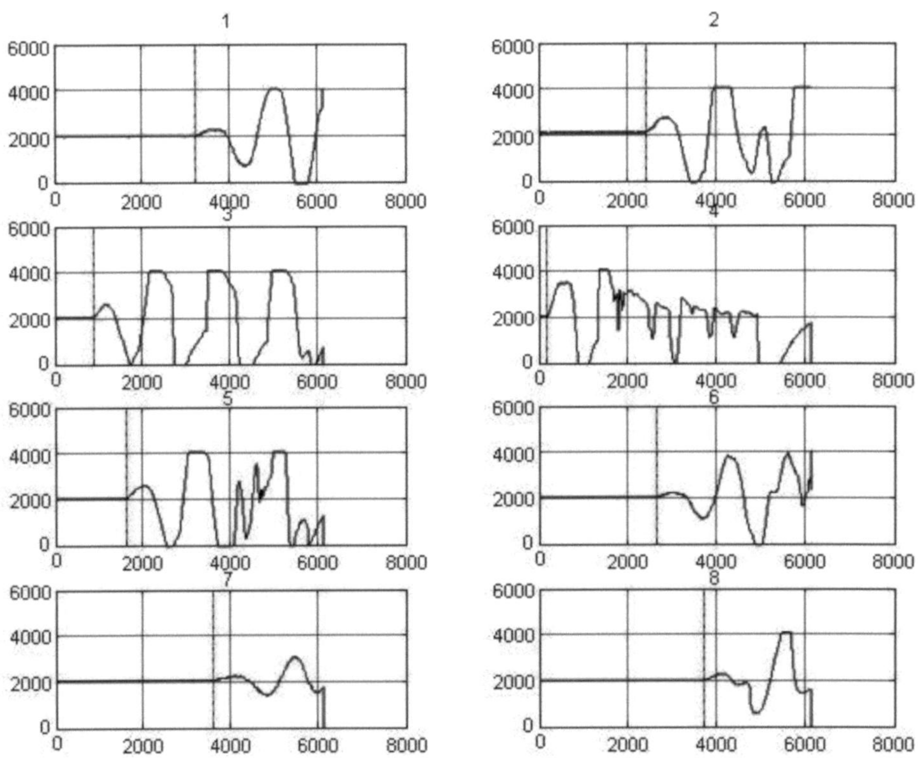

Figure 3.
An example of time realization of acoustic signals, synchronously detected by an eight-channel receiving system after a knock with a scalpel.

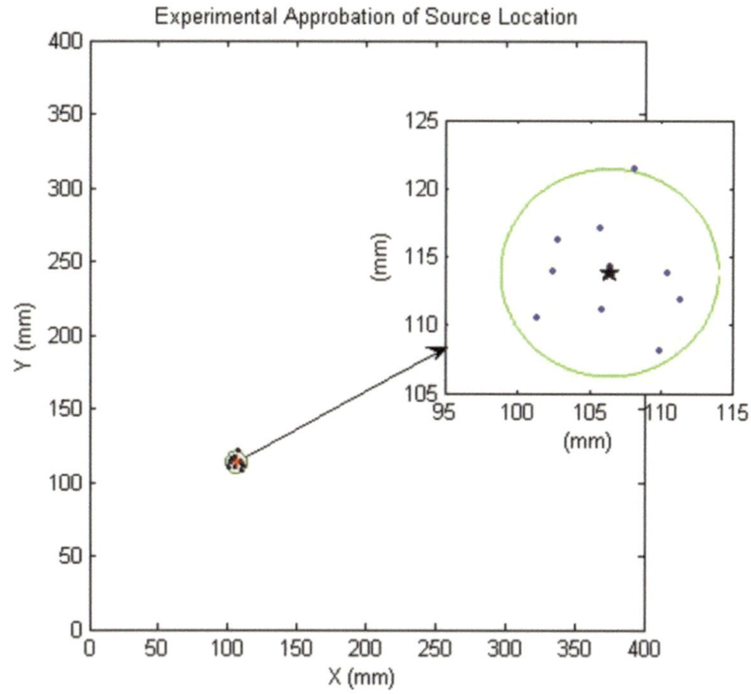

Figure 4.
The results of experimental studies of the accuracy of determining the coordinates of the source.

Figure 3 provides an example of time realizations of acoustic signals that were synchronously detected by the eight-channel recording system after fast knock on the metal plate with a scalpel. Synchronization was achieved by the electromagnetic method (i.e., the channels were triggered for recording of acoustic signals by a knock on the plate with the scalpel).

Experimental testing of the suggested method for determining the signal arrival time was performed implicitly, based on calculating the source coordinates and beginning time of acoustic signals.

A great deal of experiments were held in determining the location of signal source for different positions of the excitation point. Results of determination of source coordinates based on synchronously experimental registered realization for a knock on the plate with the scalpel near the location (110 mm, 110 mm) are presented in **Figure 4**.

6. Conclusions

The important problems of the PAT are the determination of the arrival time of the synchronously registered signals by the multichannel receiving system and calculation based on these source coordinates and the beginning time of the acoustic signal.

The method of determination arrival times of synchronously registered acoustic signals with a delayed front based on the use of an adaptive threshold is proposed. The threshold is chosen according to the linear dependence of the decreasing of threshold with the increasing of the delay.

The approach to the determination of the source coordinates and the beginning time of acoustic signal on the basis of the problem reducing to the solving of the system of linear algebraic equations is proposed, which allows to estimate the accuracy of determination of the source coordinates and the beginning time of the acoustic signals. The matrix of the system of linear algebraic equations depends on the location geometry of the source and receivers of the multichannel system. Those sources, for which the matrix of the system of linear algebraic equations is poorly conditioned, are not taken into account when collecting projective data.

The results of simulation and experimental testing of the proposed technologies confirmed their effectiveness.

Author details

Rostyslav Romanyshyn, Galyna Romanyshyn and Igor Romanyshyn*
Karpenko Physico-Mechanical Institute of the NAS of Ukraine, Lviv, Ukraine

*Address all correspondence to: romanyshyn@ipm.lviv.ua

References

[1] Sedlak P, Sikula J, Lokajicek T, Mori Y. Acoustic and electromagnetic emission as tool for crack localization. Measurement Science and Technology. 2008;**19**:1-7. DOI: 10.1088/0957-0233/19/4/045701

[2] Stepanova LN, Kanifadin KV, Ramazanov IS. The influence of temperature on the characteristics of piezoelectric transducers and errors of localization of acoustic-emission signals. Russian Journal of Nondestructive Testing. 2010;**46**(5):377-385. DOI: 10.1134/S1061830910050086

[3] Stepanova LN, Ramazanov IS, Kanifadin KV. Estimation of time-of-arrival errors of acoustic-emission signals by the threshold method. Russian Journal of Nondestructive Testing. 2009;**45**(4):273-279. DOI: 10.1134/S1061830909040081

[4] Kundu T. Acoustic source localization. Ultrasonics. 2014;**54**(1): 25-38. DOI: 10.1016/j.ultras.2013.06.009

[5] Sen N, Kundu T. A new wave front shape-based approach for acoustic source localization in an anisotropic plate without knowing its material properties. Ultrasonics. 2018;**87**:20-32. DOI: 10.1016/j.ultras.2018.01.011

[6] Yin S, Cui Z, Fu J, Kundu T. Acoustic source localization in heterogeneous media. Ultrasonics. 2019;**99**:105957. DOI: 10.1016/j.ultras.2019.105957

[7] Kalitkin NN, Yukhno LF, Kuz'mina LV. Quantitative criterion of conditioning for systems of linear algebraic equations. Mathematical Models and Computer Simulations. 2011;**23**(2):541-556. DOI: 10.1134/S207004821105009

[8] Koshovyi VV, Romanyshyn IM, Romanyshyn RI, Semak PM, Sharamaga RV. Development of passive ultrasonic tomography techniques. Russian Journal of Nondestructive Testing. 2016;**52**(10):539-553. DOI: 10.1134/S1061830916100065

[9] V. Bakul Institute for Superhard Materials. Available from: http://www.ism.kiev.ua/departmens/index?i=63

Section 2

Photoacoustic Imagine: Applications

Chapter 5

Photoacoustic Imaging in Gastroenterology: Advances and Needs

Sheena Bhushan, Sharmila Anandasabapathy and Elena Petrova

Abstract

Gastroenterologists routinely use optical imaging and ultrasound for the minimally invasive diagnosis and treatment of chronic inflammatory diseases and cancerous tumors in gastrointestinal tract and related organs. Recent advances in gastroenterological photoacoustics represent combination of multispectral and multiscale photoacoustic (PA), ultrasound (US), and near-infrared (NIR) fluorescent imaging. The novel PA endoscopic methods have been evaluated in preclinical models using catheter-based miniature probes either noncontact, all-optical, forward-viewing probe or contact, side-viewing probe combined with ultrasound (esophagus and colon). The deep-tissue PA tomography has been applied to preclinical research on targeted contrast agents (pancreatic cancer) using bench-top experimental setups. The clinical studies engaging human tissue ex vivo have been performed on endoscopic mucosal resection tissue with PA-US tomography system and intraoperative imaging of pancreatic tissue with PA and NIR fluorescence multimodality. These emerging PA methods are very promising for early cancer detection and prospective theranostics. The noninvasive transabdominal examination with PA-US handheld probe has been implemented into clinical trials for the assessment of inflammatory bowel disease. To facilitate translational and clinical research in PA imaging in gastroenterology, we discuss potential clinical impact and limitations of the proposed solutions and future needs.

Keywords: photoacoustic endoscopy, optoacoustic tomography, gastrointestinal tract, Barrett's esophagus, intraoperative, pancreatic cancer, inflammatory bowel disease

1. Introduction

Gastroenterology is a field of medicine that studies the gastrointestinal (GI) tract and its disorders. The GI tract consists of the mouth, pharynx, esophagus, stomach, small intestine (duodenum, jejunum, ileum), large intestine (also called colon which consists of the cecum, rectum, and anal canal), appendix, liver, and pancreas [1]. Imaging techniques used for the diagnosis and treatment of gastrointestinal disorders mainly include noninvasive methods (such as whole body and transabdominal imaging), minimally invasive methods (such as endoscopy), and invasive methods (such as intraoperative imaging) (**Figure 1**).

IntechOpen

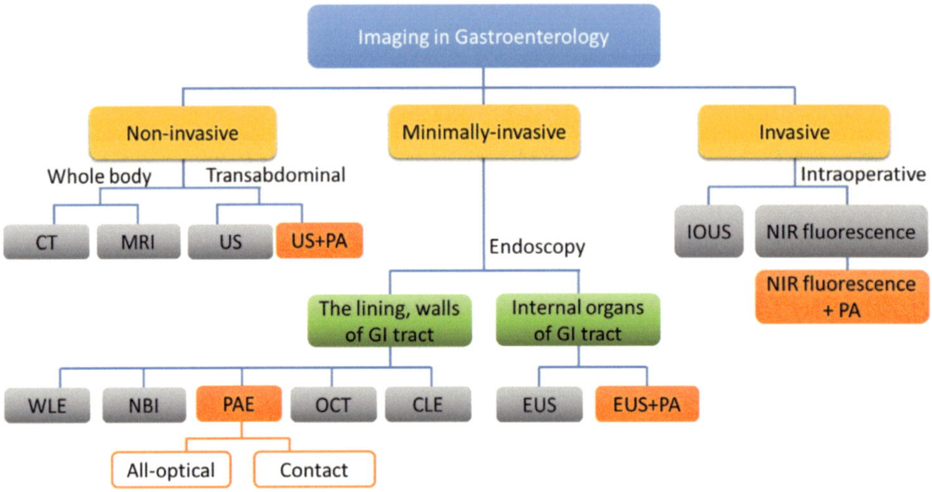

Figure 1.
Diagram of imaging modalities applied in gastroenterology: CT, computed tomography; MRI, magnetic resonance imaging; US, ultrasound imaging; PA, photoacoustic imaging; WLE, white light endoscopy; NBI, narrowband imaging; PAE, photoacoustic endoscopy; OCT, optical coherence tomography; CLE, confocal laser endomicroscopy; EUS, endoscopic ultrasound; IOUS, intraoperative ultrasound; NIR, near-infrared.

The word endoscopy is derived from the Greek words endo (within) and *skopein* (to see). Endoscopy is a medical procedure that employs a medical instrument called the endoscope, which is inserted directly into an organ or cavity (e.g., esophagus), thereby allowing a gastroenterologist to visualize and examine the organ's interior. Organs and cavities of the gastrointestinal tract that can be visualized and examined by endoscopy include (i) the esophagus, stomach, and duodenum (esophagogastroduodenoscopy); (ii) the small intestine (enteroscopy); (iii) the large intestine (colonoscopy, sigmoidoscopy); (iv) the bile duct and pancreas (endoscopic retrograde cholangiopancreatography, also known as ERCP); and (v) the rectum (rectoscopy) and anus (anoscopy); when both the rectum and anus are examined together, the procedure is called proctoscopy [2].

As a clinical specialty, gastroenterology relies upon endoscopy mainly for the following purposes:

a. *To detect* pathologies like ulcers, varices, inflammation, and bleeding—which are not visible to the naked eye.

b. *To confirm* diagnosis of cancers by taking biopsies.

c. *To treat* pathologies in a minimally invasive manner. For example, treating a bleeding vessel by cauterization, a narrow esophagus by widening (endoscopic dilation), polyps and foreign bodies by removal, and varices by banding.

Gastrointestinal endoscopy is continually evolving to enable early detection of diseases, which can significantly impact treatment outcomes. The best-known endoscopic imaging methods in practice are (1) white light endoscopy, (2) narrowband imaging, (3) optical coherence tomography, (4) confocal laser endomicroscopy, and (5) endoscopic ultrasound. In this section, we will discuss these endoscopic imaging techniques and highlight their main advantages and limitations.

1.1 White light endoscopy

White light endoscopy (WLE) is the most commonly used endoscopic imaging modality. WLE applies the full spectrum of visible white light (400–700 nm) to visualize the outermost lining of the GI tract (also called the mucosa or epithelium) [3]. Xenon lamps are used to produce white light, which is transmitted through a filter and then reflected by the mucosa. A CCD camera then processes the reflected light to an image monitor. The image can either be of a standard resolution or can be magnified up to 150 times in the more recent versions [4]. WLE detects slight color changes on the mucosa (slight redness and pallor). However, it is incredibly challenging to detect early cancer; as in early stages, the lesion is usually unremarkable, with no significant color changes. Furthermore, WLE is limited by what the human eye can see and is often associated with low sensitivity for the detection of early neoplasia. Consequently, in attempts to increase the sensitivity of WLE, chromoendoscopy is often added. Chromoendoscopy is a technique in which special stains are topically applied to the mucosa during endoscopy. Chromoendoscopy highlights subtle irregularities in the mucosa [5], and while it can help the gastroenterologist uncover some lesions invisible to the naked eye, the specificity is often quite low.

1.2 Narrowband imaging

Narrowband imaging (NBI) is one of the most widely available optical imaging techniques today. In addition to highlighting subtle differences in the epithelium (like the combination of WLE and chromoendoscopy does), NBI can also highlight changes to the subsurface mucosal vascularization, without the use of any special stains or dyes. It can be activated by pressing a button on the endoscope, and once activated it uses special filters to narrow the wavelength range of the emitted light into blue (400–430 nm; centered at 415 nm) and green light (525–555 nm; centered at 540 nm) [6]. This system exploits the principle of depth of light penetration in accordance with the Beer-Lambert law: the shorter the wavelength, the more superficial the penetration. The blue filter is designed to coincide with the peak absorption spectrum of hemoglobin, and it enhances the mucosal vasculature, like capillaries, which appear brown under NBI. The green light penetrates deeper and enhances the appearance of submucosal vasculature, like veins, which appear cyan [7]. Computer processing can further enhance the images. Since increased vasculature is one of the earliest signs associated with cancer, NBI helps with the detection of early cancer. By helping the gastroenterologist visualize and differentiate abnormal cancer margins from normal healthy tissue, it also helps determine the mucosal areas that should be biopsied. However, NBI is not able to image submucosal abnormalities.

1.3 Confocal laser endomicroscopy

The term confocal refers to the alignment of both illumination and collection systems in the same focal plane. Confocal laser endomicroscopy (CLE) focuses blue laser light (488 nm) through a single lens on to a specific target, with subsequent detection of the fluorophores (excitable using 488 nm laser light) in the tissue through a pinhole [8]. Typically, endomicroscopes have a miniaturized scanning head at the tip of the endoscope or execute the scanning outside the patient transferring the scan pattern of the tissue through an optical fiber bundle. CLE can provide high-resolution images (~0.7–1 μm) of the GI mucosa at a cellular and subcellular level, with a field of view of ~200–300 μm [9]. By delivering real-time surrogate "optical biopsies," it enables physicians to interpret histology in real time. Since the first visible neoplastic changes in epithelial cancers occur at a cellular level (such as increased nucleocytoplasmic ration, irregular shape and size of the nucleus, fragmentation of nucleus, prominent nucleoli, etc.), it facilitates early detection and treatment of cancer. It allows for targeted biopsies of abnormal mucosa, thereby decreasing unnecessary biopsy as compared to WLE. Besides the esophagus and colon, it is also able to provide an optical biopsy within the pancreas [10]. Since CLE relies upon tissue fluorescence, intravenous or topically applied contrast agents (dyes) are required [11]. The major disadvantage of CLE is a small field of view (<1 mm) when compared to the area of examined surfaces (several cm).

1.4 Optical coherence tomography

Optical coherence tomography (OCT), also known as volumetric laser endomicroscopy (VLE), is a more recent endoscopic imaging technique. OCT is based on the same echolocation principle as an ultrasound but uses light waves instead of acoustic waves [12]. The OCT imager's probe can be introduced through the accessory channel of an endoscope and be directed toward the area of interest. It is then kept in contact with the mucosal surface. OCT uses near-infrared (NIR) light (700–1500 nm range of wavelength) and directs it at the target tissue. Using an interferometer device setup, the differential light scatter is picked up, interpreted, and converted into a cross-sectional image. The typical axial resolution is around 7–10 μm which enables the cross-sectional images to provide information about the architectural morphology of the GI epithelium at a microscopic level. The lateral resolution of OCT is about 30 μm, and depending on the applied wavelength of light, depth of penetration for OCT is around 2–3 mm [13]. Since tissue is relatively transparent, the longer wavelength can penetrate deeper than CLE. However, the lateral resolution is lower than that of CLE (CLE resolution is up to 1 μm). While OCT resolution is able to detect deeper structures (e.g., crypts and glands), it cannot pick up microscopic nuclear changes, such as nuclear dysplasia. The main advantage of OCT is the deeper penetration, which would help detect cancer in the submucosal layer (the layer present right below the mucosa) [14]. On the other hand, clinical interpretation of OCT images can be challenging.

1.5 Endoscopic ultrasound

Endoscopic ultrasound (EUS) can differentiate between the various layers of the gastrointestinal tract and provide anatomical and structural details. It can employ a broad range of frequencies (5–20 MHz) to provide images with a variable depth of penetration (5–6 to 1–2 cm) and resolution (2–3 to ~0.2 mm) [15]. The most routinely used EUS array has a central frequency of 7.5 MHz. EUS is limited by a lack of specificity and an inability to differentiate between pathologies due to poor contrast (for example pancreatitis from pancreatic cancer) and is of limited use in picking up early and metastatic cancer. Additionally, EUS is a technique that largely depends on the operator's skills and has a significantly long learning curve for novice gastroenterologists [16].

The aforementioned imaging techniques have come a long way since their inception, but are still far from perfect in several ways. Imaging that relies on WLE is substantially limited by what the human eye can see and detect on the surface, with no insight into the subsurface or cellular level. Techniques like NBI and CLE are able to provide a higher resolution but cannot reach superior depths of penetration. OCT and EUS can penetrate deeper into the tissue but are limited by poor contrast, which makes interpretation of pathology difficult especially for early detection. Therefore, they are usually associated with low sensitivity and specificity. **Figure 2** summarizes the aforementioned optical methods of endoscopy.

The field of endoscopic imaging is continuously being upgraded in an attempt to address some of the abovementioned concerns. An emerging imaging technique known as photoacoustics (PA) may have a major advantage over other imaging modalities in both resolution and penetration depth.

1.6 Photoacoustic imaging

The basic principle of photoacoustic (aka optoacoustics) imaging is derived from the photoacoustic effect: (1) it uses a pulsed nanosecond-long laser light to illuminate a target tissue, and (2) when the energy of laser light is rapidly absorbed and followed by fast heating of material, it is converted into pressure wave via thermoelastic expansion of the targeted tissue. Using detected acoustic signals, the PA images can be reconstructed by means of algorithms of computed tomography (CT).

The amplitude of a generated PA pressure wave, p, can be described by the formula [17, 18]:

$$p = \mu_a F \Gamma = \mu_a F \frac{\beta V_l^2}{C_p}, \tag{1}$$

with the optical absorption coefficient μ_a, the local optical fluence F, and the dimensionless Grüneisen parameter Γ representing thermoelastic efficiency of the medium. The Grüneisen parameter is temperature sensitive and depends on the volumetric thermal expansion coefficient (β), the speed of sound (V_l) for longitudinal waves, and the specific heat capacity at constant pressure (C_p) [19, 20]. In the tissue, a rise of temperature $\Delta T \sim 10^{-3}$°C generates a pressure wave of p ~ 1 kPa.

The depth of imaging is correlated with (a) the depth of light penetration in materials and (b) the detection sensitivity of an US transducer that varies with its frequency range. Since PA imaging combines the use of light and sound, it can provide spatial resolution at the scale of optical imaging (up to submillimeter range) and a penetration depth across US dimensions (at centimeter range).

	Endoscopy		Endomicroscopy	
	WLE	NBI	CLE	OCT
Resolution	~100 μm	<100 μm	~0.7-1 μm	~7-10 μm
Depth	~10-30 μm	~300-400 μm	~300 μm	~2-3 mm
Field of view	140°	140°	~200-300 μm	~5-6 mm
Objectives	Tissue surface	Subsurface	Cellular	Architectural
Figure				

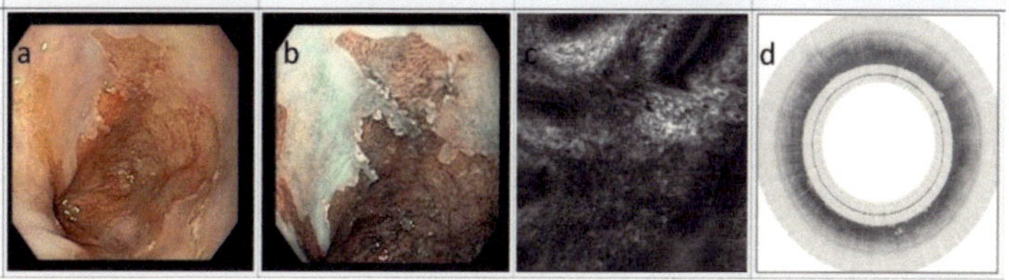

Figure 2.
Comparison of endoscopic imaging modalities: (a) WLE, (b) NBI, (c) CLE, and (d) OCT. (a–c) Images by Prof. Sharmila Anandasabapathy, MD. (d) Image courtesy of Prof. Guillermo J. Tearney, MD, PhD.

Resolution of PA images can be affected by multiple factors. Heat-generating optical energy must be delivered to the tissue faster than the producing pressure wave can propagate the distance equal to the desired spatial resolution. Therefore, the use of nanosecond optical pulses is a necessary (but not sufficient) condition to achieve the desirable spatial resolution. Besides, a temporal response function of a pressure-wave detector has to be not slower than the optical-pulse duration to make sure that the desirable spatial resolution is achievable.

Tissue structure can be represented by PA images only if acoustic detectors are capable to resolve rapid changes in PA signals associated with boundaries and sharp edges in tissues and reproduce slow changes associated with smooth variation in optical properties within the tissue at once. That is to say, acoustic detectors must register both high and low ultrasonic frequencies of acoustic pressure at the same time. Such acoustic detectors are called ultra-wideband acoustic transducers [18, 21].

It is challenging to detect PA signal properly. The US detection bandwidth of an acoustic transducer defines the limits of axial (depth) resolution. Furthermore, the lateral resolution of PA images is provided by an array of transducers and depends on the dimensions of each acoustic transducer and the geometry and dimensions and of the US transducers in an array. Scanning a single transducer along tissue surface can simulate the array.

In practice, PA imaging is frequently combined with conventional US detectors. At the same time, conventional US detectors have relatively narrowband characteristics due to its design to both transmit and receive acoustic wave. Spectral amplitudes of typical PA signal and US transducer parameters are illustrated on **Figure 3**.

Conventional piezoelectric transducers limit the potential axial resolution of broadband PA signals. Note that all-optical detectors of PA signal are wideband and have significant advantage in these terms [22, 23].

In general, the depth-resolution ratio defines the quality of PA imaging system. The tradeoff between depth and resolution leads to three types of PA imaging methods: microscopy (μm-scale, cellular), mesoscopy (mm-scale, microvasculature), and macroscopy (cm-scale, organs).

Scanning PA microscopy can demonstrate resolutions greater than 50 μm (e.g., resolving capillary bed and even single erythrocytes) [24] at depths that are beyond the optical diffusion limit, something which is not achievable with conventional microscopy methods (depths are up to 4–5 mm). Additionally, spatial resolution and penetration depth are somewhat mutually exclusive in imaging techniques like the ultrasound (i.e., as the penetration depth increases, spatial resolution decreases and vice versa). PA imaging in gastroenterology mostly employs meso- and

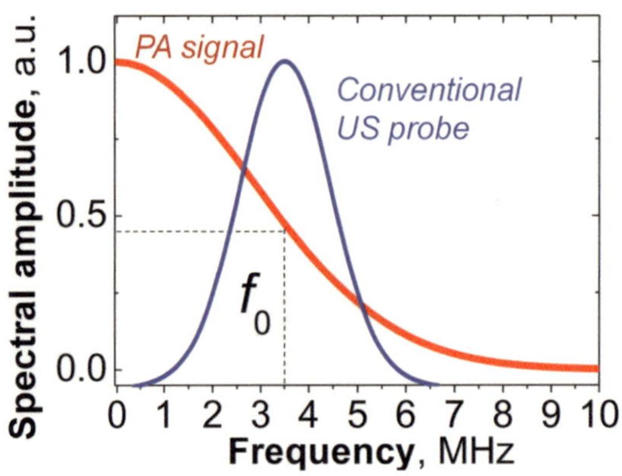

Figure 3.
Spectral amplitudes of broadband PA signal and conventional US transducer.

macroscopic scale. PA deep-tissue imaging uses NIR light to generate an acoustic signal allowing for more clinically relevant penetration depth up to several centimeters, along with maintenance of spatial resolution (resolution of 0.2–2 mm at reported imaging depths of 3–7 cm) [18]. Combining clinically relevant depth with high resolution can make it possible to unmask inconspicuous "flat" lesions during early stages of cancer development. It can also help with determining the extent of the tumor depth both intraoperatively and postoperatively. **Figure 4** resumes the depth of penetration associated with the aforementioned imaging modalities on the example of the esophageal lining tissue.

Unlike other imaging modalities, PAI can use endogenous chromophores such as oxyhemoglobin, deoxyhemoglobin, lipid, water, and melanin. The NIR light has longer wavelengths (700–1000 nm) than the visible light range, and these wavelengths can be used to reach deep within the tissue and provide detailed information about its composition. In addition to revealing normal physiological characteristics (total hemoglobin concentration, O_2 saturation of hemoglobin, blood flow, temperature), it can help physicians detect hypoxia, perfusion, and increased vasculature (also called neoangiogenesis—which is an early sign of cancer) through *anatomical* and *functional* imaging, relying exclusively on endogenous contrast agent (without dyes). Exogenous agents can also be used to provide *molecular* imaging, to enhance the contrast further, and to facilitate deeper penetration. They can be used in a targeted manner to target molecular-specific processes, such as biomarkers specific to inflammation or certain type of tumors. For example, cancer development is associated with increased permeability of the vasculature, which leads to extravasation of the contrast agent use. Overall, by providing a high optical contrast, it helps differentiate the cancer margin from the surrounding healthy tissue in real time. Lastly, it is quite similar and also naturally compatible with the ultrasound in

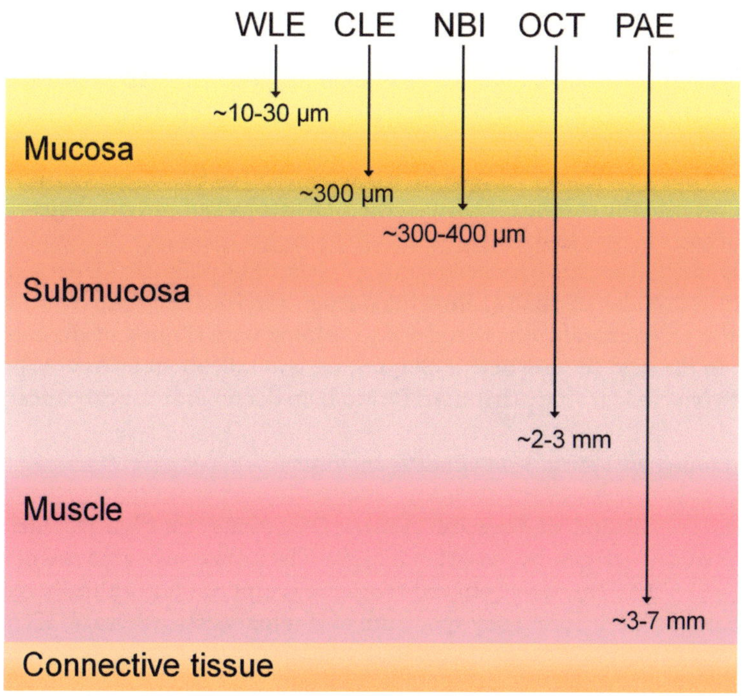

Figure 4.
Schematic summarizing the depths of penetration into esophagus tissue for routine and emerging imaging techniques.

principle and operation. It is portable and can be used in real time and requires no special training. It does not utilize ionizing radiations and is therefore minimally invasive and safe, making it an easily adoptable technology for physicians. In the upcoming sections of this chapter, we will discuss most recent PA imaging applications for gastroenterology.

2. Applications of photoacoustic imaging in endoscopy

2.1 Esophageal cancer

Esophageal adenocarcinoma carries a significant disease burden in the Western world. Barrett's esophagus (BE) is the most important risk factor for developing esophageal adenocarcinoma. Patients with BE are at 30- to 125-fold increased risk of developing cancer than the general population [25]. Patients who develop high-grade dysplasia (HGD) from BE have an even higher chance of developing esophageal adenocarcinoma [26]. Indeed, early detection of Barrett's esophagus significantly improves treatment outcomes, yet accurate detection of dysplasia in BE still remains a challenge. In BE, the normal esophageal squamous mucosa (which appears white under WLE) is replaced by an intestinal-type columnar mucosa (appears salmon pink under WLE). The current standard of care includes taking targeted biopsies of visible lesions under WLE guidance, which are then sent over to histopathology for staging. However, since this approach cannot sample the entire esophagus, the chances of missing dysplasia (precursor to cancer) are very high [27]. Additionally, the screening is just limited to the mucosa, making detection of submucosal invasion impossible.

Indeed, techniques like NBI and CLE yield high-contrast, subcellular images, but they fall short in terms of depth (1.5–2 mm) [28], which is not sufficient to detect submucosal invasion. Tumors that have invaded the submucosa are treated by surgical removal of the esophagus (esophagectomy) which is often associated with significant mortality and decreased quality of life, whereas tumors that have not yet invaded the submucosa can be treated by less aggressive techniques like endoscopic mucosal resection (EMR). There is a clinical need for developing an imaging modality that can detect dysplasia in vivo and also image beyond the mucosa layer into the submucosa. Since PA microscopy can provide high-resolution imaging of the microvasculature, it can help to bridge these gaps by helping physicians to differentiate between nondysplastic and dysplastic BE early on. Additionally, it is also capable of imaging deeper into the tissue (up to 3–5 mm) and reaching the submucosal layer, thereby also helping with staging of the cancer in real time. In the next section of this chapter, we will talk about an in vivo preclinical and an ex vivo clinical study that used PAI to help detect esophageal cancer.

2.1.1 Noncontact, all-optical photoacoustic endoscopy

A miniature, all-optical, forward-viewing PA catheter for high-resolution 3D endoscopy was developed by Ansari et al. [29]. The sensor provided the necessary broadband (1–70 MHz), well-behaved frequency response for faithfully recording the wideband acoustic frequency spectrum of photoacoustic waves. A 3D image had a cylindrical field of view of 3.5 × 7 mm; the diameter of the catheter was 3.2 mm. The lateral resolution ranged from 40 μm at a depth of 1 mm to 175 μm at a depth of 7 mm. These variations arise because the solid angle that is subtended by the absorbers to the FP sensor plane decreases as the depth increases. The first tests were performed on highly vascularized tissue samples and demonstrated great potential of the method (**Figure 5**). Excitation laser wavelength, 590 nm; pulse repetition frequency, 30 Hz; and incident fluence, 15 mJ/cm^2, which is lower than the American National Standards Institute (ANSI) laser safety limit (20 mJ/cm^2), were used.

The microvasculature can be seen very well. The data acquisition time was as long as 25 min and cannot be considered as clinically relevant; though, in the future this parameter can be improved applying a faster laser with pulse repetition rate up to 200 Hz and parallelizing readout of the sensor. The technique described above addressed some of the needs for early diagnostics of esophageal cancer (since BE is associated with increased vasculature) and demonstrate capability to image partially submucosa. However, to the best of our knowledge, no examination of this method was performed on esophagus so far either in vivo or ex vivo.

2.1.2 Catheter-based photoacoustic ultrasound endoscopy

The use of contact (coupling required) multimodal photoacoustic and ultrasonic endoscopic imaging to two rabbit esophagi, to investigate the esophageal lumen structure and microvasculature, was demonstrated by Yang et al. [30]. Both in vivo and ex vivo studies were performed, and high-resolution photoacoustic images of vasculature network in the walls of GI tract were acquired and analyzed. To perform this experiment, they used a catheter-like side-viewing photoacoustic endoscope/endoscope ultrasound miniprobe (PAE-EU). This setup allowed to image internal walls of the GI tract using both PA and US simultaneously. A wavelength of 584 nm with a laser energy of 0.3 mJ/pulse was used to produce PA image contrast proportional to the hemoglobin concentration. The detection transducer was fabricated as ~36 MHz, 65% fractional bandwidth that allowed a penetration depth of up to 3.5 mm. The field of view of the endoscopic scanning system was 7 mm, and its angular field of view was set to 270 degrees. A focused US transducer detected 1D depth-resolved signals (or A-lines), and cross-sectional images (or B-scans) were produced using rotation of a scanning mirror that directed optical and acoustic waves. Data acquisition system collected dual-wavelength photoacoustic and

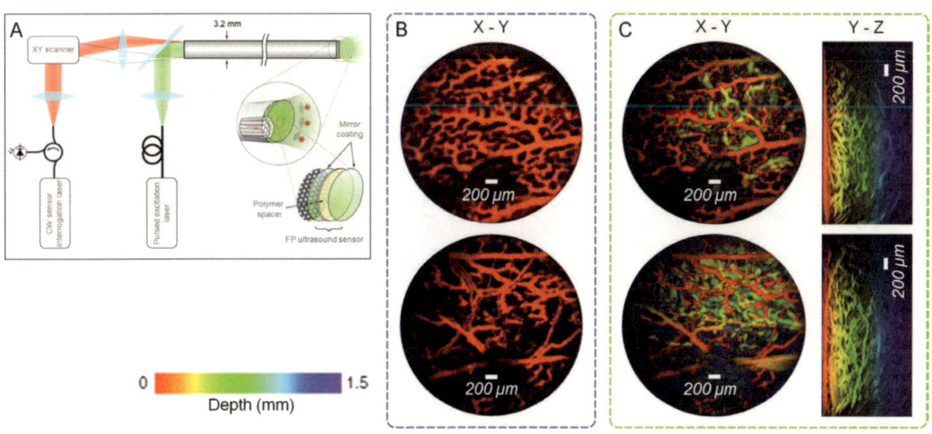

Figure 5.
(A) Schematic of the all-optical forward-viewing PA endoscopy probe includes the magnified distal end of individual fiber-optic cores in the coherent fiber bundle and ultrasound sensor based on the principle of fabry-perot (FP) interferometer. (B) High-resolution PA images of ex vivo capillary network of the chorioallantoic membrane that surrounds the avian embryo. X-Y maximum intensity projections (MIPs) for depth range z = 0–200 μm of two regions on the same sample with the microvascular anatomy. (C) X-Y MIPs for the depth range z = 0–1.5 mm for the same two regions as Panel B, respectively, and Y-Z MIPs. The images are reproduced from open-source publication [29] with author permission.

ultrasonic B-scan images in real time (~4 Hz) during the experimental procedure. The endoscopic probe had a diameter of 3.8 mm and a working distance of 0.5 mm from the probe surface which were optimal for a rabbit esophagi. The recently upgraded version has diameter of 3.2 mm and fits within the 3.7 mm diameter of instrument channel of a standard clinical video endoscope (**Figure 6**).

Two intact rabbit esophagi were imaged using the above setup. The rabbits were fasted for 12 h and adequately anesthetized before the procedure. They were then laid down at a slight incline in a supine position. Since water is able to provide acoustic couple, the esophagi were filled with water; following which the endoscopic probe was inserted through the mouth and advanced using gentle force, until it could no longer be advanced gently (25–30 cm). At this point, simultaneous PA and US imaging were initiated and images were acquired. Following in vivo imaging, the rabbits were euthanized, and their esophagi were imaged again in an ex vivo setting, using the same procedure. On completion of the ex vivo imaging, samples were collected for histology, and the results were later compared with the PA and US images.

In vivo imaging results: cross-sectional PA images and corresponding coregistered US images of vasculature within the esophagi were acquired. Since the images were acquired over several respiration cycles, they had significant motion artifact, which resulted in an uneven mapping of the vasculature. With assistance from a filtering algorithm and at the expense of spatial resolution, the motion artifacts were artificially corrected. Ultimately, only larger vascular structures and organs like

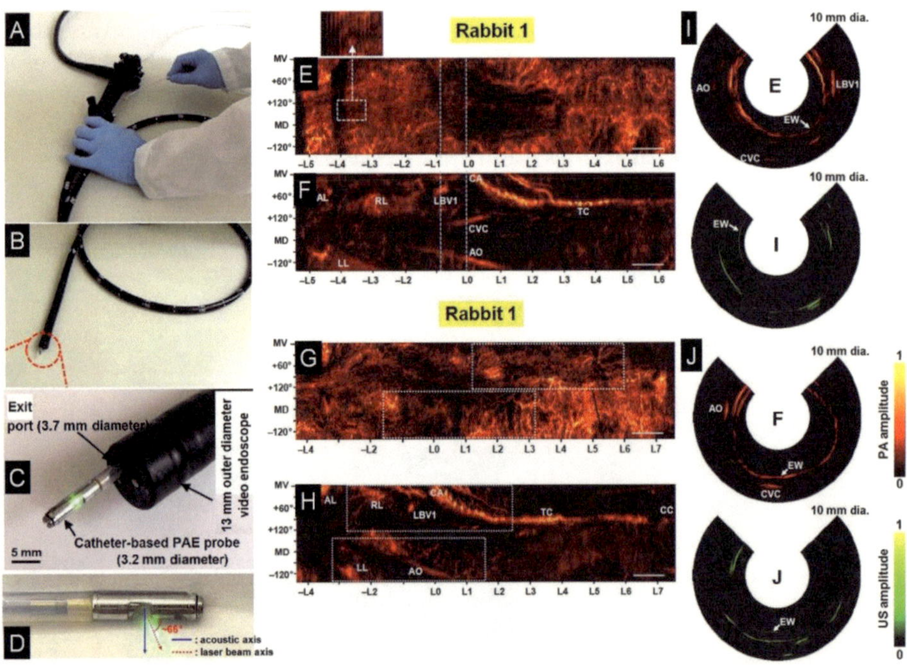

Figure 6.
(A and B) Photos of the catheter-based PA and US probe within the 3.7 mm channel of a clinical video endoscope. (C and D) Side-viewing design of distal end of the combined probe. (E and G) Normalized PA images of rabbit esophagus show the total Hb distribution in vivo and ex vivo, correspondingly. The left side is the lower esophagus; the right side is the upper esophagus. (F and H) Total Hb distribution in the mediastina in vivo and ex vivo, correspondingly. AO, aorta; TC, trachea; AL, accessory lobe; LL, left lobe; RL, right lobe; CA, carina; CVC, caudal vena cava; LBV1, 2, large blood vessels; CC, cricoid cartilage. Scale bar, 10 mm. (I and J) B-scan images PA (red) and US (green) as indicated at E, F: esophageal wall (EW) is less contrasted on US image than on PA image. The images are reproduced from open-source publication [30] with author permission.

the inferior vena cava, aorta, and parts of the pulmonary vasculature were visible. Unfortunately, following the correction, the finer vessels were blurred out and were not visible.

Ex vivo imaging results: the absence of motion artifact in the ex vivo experiments enabled the acquisition of higher-quality, uninterrupted 3-D maps of the esophageal vasculature. The PA data was clearer, stronger, more detailed, and overall superior than the US data. Vascular structures (both major and feeding vessels) were clearly visible, whereas in US the smaller vessels were not as clear. Using PA, the group was also able to measure the blood vessel diameter, based on the PA esophagus distance maps. On comparing PA and US images with histology, it was found that the PA images provided clear boundaries of the esophageal wall, consistently throughout the length of the esophagus. This can be attributed to a strong PA signal generated by the capillary dense submucosal layer. However, the US did not show clear boundaries of the mucosal and submucosal layers consistently, mainly due to weak signal intensities in these layers. Wall thickness of the esophagus calculated using PA varied only slightly from the histological images (PA were thinner than histological images). This slight change can be attributed to distortion of the tissue during histological fixation. Thorough analysis of PA-US overlapped images made it possible to identify the submucosal layer of the esophagus along with the surrounding vasculature like the aorta and inferior vena cava [30]. Thus, PAE can provide label-free visualization of microvasculature (as small as 190 μm), which exceeds the capability of current EUS technique in the esophagus. While Doppler ultrasonography can image blood vessels, it can only image large blood vessels which lowers its sensitivity. Since PAE can image smaller blood vessels, it is far more sensitive and could prove to be an invaluable asset for diagnosing BE. Moreover, PAE can also visualize vasculature in the surrounding mediastinal region, which can prove helpful intraoperatively (during esophagectomy). Since a majority of modern endoscopes have narrow working channels, further miniaturization of the tool to about 2.8 mm in diameter is required. Furthermore, if motion artifacts can be decreased, this technique can be used not only for diagnosing BE but also for guiding surgical procedures like esophagectomy. Motion artifacts can be reduced by increasing the scanning speed and by adding respiratory gating to the image acquisition. Next, we will talk about an ex vivo study that was done in human samples.

2.1.3 Ex vivo endoscopic resection tissue with PA tomography system

An ex vivo feasibility study in humans that used PA imaging in patients undergoing EMR for dysplasia was conducted by Lim et al. [31]. The aim of this study was to identify microvascular patterns associated with dysplasia in ex vivo samples, with the long-term goal of performing in vivo PA imaging for the detection of dysplasia in patients. In this study, a commercially available photoacoustic tomography imaging system Vevo LAZR (VisualSonics, Toronto, Ontario, Canada) was applied. A linear-array US transducer of 40 MHz central acoustic frequency and 256 elements (LZ550, VisualSonics) were integrated with fiber-optical bundles to provide a cross-laser-beam geometry for optical excitation. Photoacoustic and US images were recorded and displayed at a frame rate of 5 Hz. PA and US scans were obtained from a total of 13 ex vivo EMR samples from 8 patients. Excised tissues were mounted using ultrasound gel onto a plastic dish filled with gelatin (5% weight of 40 mL volume) to minimize acoustic signals from the dish. Because of the cross-laser-beam geometry of the PA imager fiber bundle attached to the transducer, there was an optimal photoacoustic zone located between 9 and 11 mm in front of the probe. Hence, they positioned the top surface of the tissue at 9 mm for consistency between samples.

PA imaging scans were made at each wavelength of 680, 750, and 850 nm. Additional measurements were made at 824 and 970 nm in anticipation of future work, in which this data would serve as controls in studies of topical biomarker-targeted porphyrin-lipid nanoparticle contrast. The transducer was mounted onto a motorized translational stage and scanned over a 3D volume at ~0.06 mm step size. The PA imaging laser power was below the ANSI maximum permissible exposure for skin (20–70 mJ/cm2 for 680–970 nm), avoiding any tissue heating that might damage the tissue, as confirmed on subsequent histopathology. Once data acquisition was completed, the specimen was marked and sent for histopathology.

The relative concentrations of oxyhemoglobin (HbO_2) and deoxyhemoglobin (Hb) were estimated by PA imaging intensity using two wavelengths of 750 nm and 850 nm. The relative total hemoglobin was then given by the sum of relative HbO_2 and Hb values. PA images were used for a 3D reconstruct of the complete EMR tissue. The US provided structural information, while the PA imaging provided functional information about the distribution of endogenous absorbers (blood) within the tissue. While there were no differences seen relative hemoglobin (HbT) among the various subclasses of BE, a higher relative HBt was seen in BE as compared to normal squamous mucosa. Since BE is associated with increased vasculature, it makes sense that it appears salmon pink under WLE (whereas normal tissue appears white). Even though this study did not utilize submucosal tissue, PA imaging data was detected beyond 2 mm from the tissue surface. A major limitation of this study was that distinction between dysplastic BE and nondysplastic BE could not be made on resected tissue. Further studies are needed to find out if PA imaging can help detect dysplasia. However, owing to its penetration depth capability, PA imaging can be very useful for endoscopic cancer staging. Note that the selection of more optimal PA imaging system also can improve the outcome.

2.2 Inflammatory bowel diseases

Inflammatory bowel disease (IBD) is widely prevalent and carries a significant burden around the world [32]. While the diagnosis is usually a clinical one, colonoscopy is performed for confirmation. Patients with a long-standing history of the disease (8–10 years) are at a higher risk of developing colorectal cancer. Annual screening colonoscopies are advised to detect early cancer changes (dysplasia) and to monitor the status of the disease in these patients. Colonoscopy is a minimally invasive but uncomfortable procedure that is necessary in this patient population. Real-time multispectral optoacoustic tomography (MSOT) imagers can detect the oxygenated and deoxygenated hemoglobin in tissue, thereby allowing physicians to detect structural and vascular changes associated with IBD in a noninvasive manner. More recently, MSOT can use combined photoacoustic and ultrasound signals, which further improves its radiologic capabilities. In this section we will discuss three studies that used MSOT to study IBD. First, we will talk about a preclinical study that used MSOT to study inflammatory and early cancer changes associated with IBD, in mice.

2.2.1 Preclinical studies

Bhutiani et al. [33] tested the usefulness of this MSOT to assess inflammatory and dysplastic changes associated with IBD, in bowel murine models. The goal of the study was to assess diagnostic capability of PA imaging of endogenous contrast agent (hemoglobin) for inflammation of the GI tract organs in preclinical model. A total of 9 mice were imaged from the thorax to the pelvis using the preclinical MSOT system inVision 256 TF (iThera Medical). Wavelengths of 680, 710, 730,

740, 760, 770, 780, 800, 850, and 900 with an acquisition time of 10 ms per frame were used for imaging. Imaging was done at three time points, the first one was before bacterial inoculation, the second one was done 2 days post bacterial inoculation, and the final imaging was done 7 days post bacterial inoculation. The images were reconstructed through back projection with 75 μm resolution. Following imaging, colonoscopy was performed on the mice, and a Colitis Severity score was obtained using a standard severity index. After colonoscopy, the mice were euthanized, and samples from their colons were taken and sent for histopathology.

The mice in the experimental group, who had bacterial-induced colitis, showed an increased mesenteric and colonic vascularity. An increase in the mean signal intensity of oxygenated hemoglobin by MSOT was also observed in the experimental group as compared to the control group. The intensity of the signal was mild at time point two (i.e., 2 days post bacterial inoculation) and was most remarkable at point three (i.e., 7 days post bacterial inoculation). The deoxygenated signal obtained through MSOT showed no difference between the two groups at any time point. The findings on MSOT were confirmed by the colonoscopy findings. The mice in the experimental group, who had bacterial-induced colitis, had higher colitis score than the control group at time point two (2.5 vs. 1.5). At time point three, i.e., 7 days post bacterial inoculation, the experimental group had an average score of 5.5 with visible vascular and structural changes. Both the MSOT and colonoscopy findings were also confirmed by histopathology. Mice in the experimental group showed increased infiltration by inflammatory cells along with architectural destruction, at both time points two and three. We will now discuss two clinical models that used MSOT to assess changes associated with Crohn's disease in humans.

2.2.2 Clinical studies

Using clinical, handheld probe MSOT, intestinal inflammation in 91 patients with Crohn's disease (ClinicalTrials.gov number, NCT02622139) was detected transabdominally as reported by Knieling et al. [34]. MSOT was used to assess hemoglobin levels in the intestinal walls of patients with active and non-active Crohn's disease. Using clinical, endoscopic, and histological scorings, patients were classified into two groups: patients who had active Crohn's disease (inflammation) and patients with no active disease (no inflammation). Ultrasonography was also done, following which the patients underwent MSOT imaging. Comparisons were made between the distributions of MSOT measurements between the two groups (patients with active Crohn's vs. non-active disease). The diagnostic performance of MSOT was also compared with that of ultrasonography. It was found that there was a significant difference in single wavelength measurements at 760 nm and spectrally unmixed total hemoglobin between patients in remission and those with active disease. While this study was limited by a small sample size, it suggests that MSOT can be used to differentiate patients in remission from those with the active disease. If patients are in remission, they do not need more invasive procedures (**Figure** 7).

In another set of experiments conducted by Waldner et al. [35], as a part of the same clinical trial (ClinicalTrials.gov number, NCT02622139), MSOT was used as rapid, noninvasive method to assess disease activity in patients with Crohn's disease. CD activity was determined based on clinical, endoscopic, and histologic evaluation along with ultrasonography. A handheld MSOT device (3–4 MHz, 256 transducer) was placed on the abdominal wall and used to provide images of the intraabdominal organs. B-mode ultrasound imaging was used in conjunction to localize parts of the intestine for accurate photoacoustic measurements.

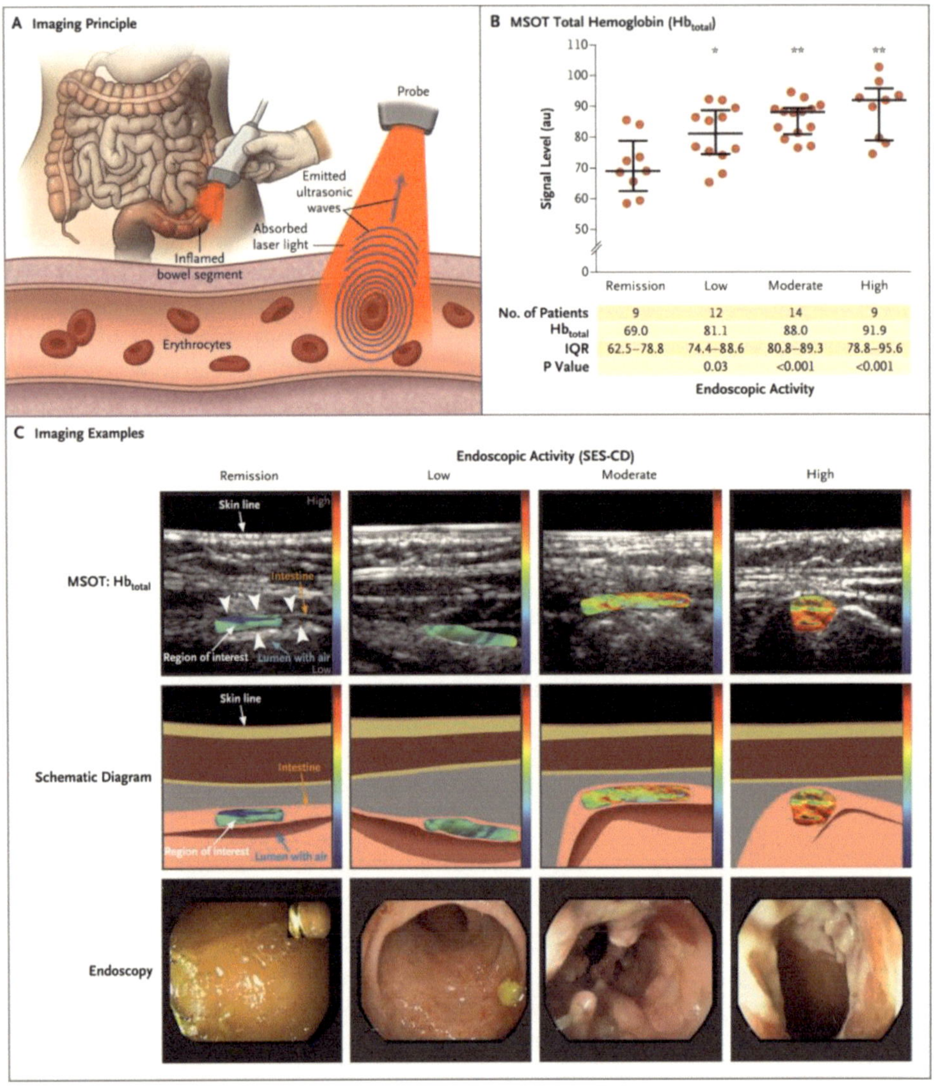

Figure 7.
Multispectral photoacoustic tomography for the assessment of Crohn's disease activity. (A) Transabdominal imaging method using handheld US probe upgraded with laser light illuminators. PA data were acquired at the wavelengths of 700, 730, 760, 800, 850, and 900 nm. MSOT system calculated total hemoglobin (Hb_{total}), oxygenated hemoglobin, deoxygenated hemoglobin, and oxygen saturation to evaluate tissue perfusion and oxygenation as factors of inflammation. (B) Normalized signal level indicated Hb_{total} in the intestinal wall of both the large bowel and the small intestine as a function of endoscopic inflammation degree. Simplified endoscopic score for Crohn's disease (0–56) was applied to evaluate severity of intestinal inflammation: Remission <3; low disease activity 3–6; moderate disease activity 7–15; high disease activity >16. IQR is the interquartile range; a single asterisk points to $P < 0.05$ and a double one to $P < 0.001$. (C) The top row represents pseudocolored maps of Hb_{total} in the large bowel and small intestine with an overlay of US images. The middle row exhibits schematics of the top row images. The bottom row demonstrates the corresponding endoscopic images used for evaluation of inflammation. The images are reproduced from [34]. Copyright © 2017 Massachusetts medical society. All rights reserved.

MSOT signals were acquired with 700-, 730-, 760-, 800-, 850-, and 900-nm excitation wavelengths. Results showed that Hb_{total}, HbO_2, and SO_2 continuously increased and correlated well with disease activity. Since MSOT signals can be obtained within 5 minutes, this technique can be used as a rapid, noninvasive way to assess disease activity in patients with Crohn's disease.

MSOT offers many advantages over the current imaging modalities. It is portable and noninvasive with a high depth of tissue penetration (up to 5 cm). It does not require contrast agents and quantifies oxygenated hemoglobin, to inform physicians about inflammatory and vascular changes associated with IBD. Therefore, it can be used as a noninvasive objective assessment tool not only to diagnose IBD but to also monitor its severity, progression, and transformation into cancer. Small changes in MSOT imaging can prompt physicians to alter treatment course and possibly also help with early detection of cancer. Finally, since it is noninvasive and less uncomfortable than a colonoscopy, it is also likely to be accepted by patients.

2.3 Pancreatic cancer

Pancreatic cancer is the seventh leading cause of cancer-related mortality worldwide [36]. The 5-year survival rates are <7%, mostly due to a delayed diagnosis and ineffective treatment methods [37]. Treatment is usually done by surgical resection of the tumor, and only 20% of patients are eligible for resection at their initial diagnosis, due to locoregional spread and metastasis [38]. The 5-year survival rates can be significantly improved if the diagnosis is made at an early stage (no metastasis) and if the patient has cancer-free margins post-surgery [39, 40]. When the diagnosis is made at an early stage, the tumor is usually small, well defined, and localized. During surgical resection of the tumor, the surgeon aims to remove the entire cancerous tissue (to get cancer-free margins); however, margin-positive resections occur very frequently (up to 70% of PDAC cases) [41]. Due to the growth pattern of the tumor and inability to differentiate between normal and cancerous tissue, small tumor extensions are usually missed during surgery. This increases the chances of metastasis post-surgery which impacts long-term survival rates negatively.

The current imaging modalities used for diagnosis and treatment are X-ray radiography, transabdominal ultrasound, endoscopic ultrasound, computed tomography, magnetic resonance imaging (MRI), positron emission tomography (PET), and single-photon emission computed tomography (SPECT). While sensitivity for detecting tumors larger than 2 cm is very high with modalities like CT, the sensitivity for tumors less than 2 cm is only 77% [42]. Imaging techniques like ultrasound, CT, and MRI detect structural and anatomical changes in the pancreas. They cannot detect molecular changes, and since it takes years of molecular changes for the structural changes to manifest, these imaging techniques have low specificity for early detection of pancreatic cancer. Molecular imaging techniques like PET and SPECT offer more sensitivity and specificity. But PET and SPECT involve ionizing radiation and thus are harmful to living organisms. Another major limitation of the conventional imaging techniques is that most of them cannot be employed as intraoperative imaging tools [43].

The only tools currently available intraoperatively to assist the surgeon with resection are intraoperative ultrasound (IOUS) and intraoperative frozen section analysis (IFSA) by a pathologist. IOUS helps the surgeon to anatomically differentiate cancerous tissue from the surrounding structures [44, 45]. It provides high-resolution, real-time intraoperative imaging and helps determine resectability of the tumor [44]. However, it is not as reliable for detecting smaller superficial lesions [46]. Additionally, it requires considerable amount of training and experience both for operation and for interpretation [47]. IFSA also helps determine resectability of the tumor, when locoregional spread is identified. It helps the surgeon to ensure negative margins after resection [48]. However, it is time-consuming and associated with high false-negative results [49, 50]. The inability to accurately identify tumors intraoperatively can lead to incomplete resections, which results in metastasis post-surgery. Additionally, patients with undetectable micrometastasis, who are not eligible for resection, may undergo unnecessary surgeries without any impact on the outcome. Thus, there is a need for newer imaging techniques that can help with early detection pf pancreatic cancer and also help stratify patients better for treatment. Since PA imaging combines the use of light and sound waves, it can provide real-time, high-resolution, functional, and molecular imaging without ionizing radiation—a significant advantage over the other imaging techniques like CT, PET, and US. As previously discussed, PA imaging also allows penetration up to clinically relevant depths, thereby making it a favorable imaging tool that can be used intraoperatively. Additionally, when used in conjunction with tumor-targeting molecular agents, it has the potential to provide crucial information to the surgeon during intraoperative use. Lastly, PA imaging is very complimentary both to US imaging and fluorescence imaging and can be easily adopted by gastroenterologists.

2.3.1 Preclinical studies on diagnostics

Using exogenous agents in PA imaging can further enhance its diagnostic capabilities. Homan et al. [51] developed a PA contrast agent—silver-silica nanocages (180–520 nm in diameter), to enhance PA imaging in pancreatic tissue. The nanocages were injected into the ex vivo porcine pancreas which were then imaged using a combined PA and US benchtop experimental setup (PA-US). The PA-US used an 800 nm light with a 7 ns pulse duration at a 10 Hz repetition rate. Optical fibers were symmetrically bundled on two sides of the US linear transducer array with 7 MHz center frequency, 14 mm wide, and 128 channels, and this arrangement allowed to produce PA and US images along the same plane. The imaging results showed that nanocage contrast agents enhanced PA imaging contrast.

Further, Homan et al. [52] also used silver nanoplates as PA imaging contrast agent to conduct in vivo PA and US imaging in transgenic mice with orthotopically grown pancreatic cancer. At first, PA and US images of the tumor were obtained without the nanoplates. Subsequently, silver nanoplates were injected into the tail vein, and imaging was done every hour following the injection, up to a total of 6 h. PA signals were captured in the wavelengths of 740–940 nm to differentiate the signals from blood and nanoplates using known optical spectra. A 3-D reconstruction of the orthoptic tumor was made, and a heterogenous distribution of nanoparticle accumulation in the tumor was noted. These findings were confirmed on histology. Because silver nanoparticles accumulate in the tumor and can provide contrast-rich images of the tumor using PA imaging, they have tremendous potential particularly in studying the molecular profile of pancreatic cancer.

Dai et al. [53] conducted a study in which they used in vivo murine models to obtain molecular images of pancreatic tumor at transmission (forward projection) mode of PA imaging. They used multimodal endoscopic approach combined with targeted multifunctional iron oxide nanoparticles (IONPs). The authors evaluated the efficiency of the proposed biomarkers to identify margins of pancreatic tumor using contrasted PA imaging and compared to NIR fluorescence. Results showed that the contrast between the tumor and normal tissue was very low to identify the tumor with nontargeted molecules (NIR830-IONP), whereas tissue with targeted molecules (NIR830-ATF-PEG-IONP) showed greater contrast between normal and cancerous tissue. Additionally, the tumor margins were clearly demarcated. The use of targeted multifunctional nanoparticles significantly improved the signals of the tumor.

2.3.2 Clinical studies on intraoperative photoacoustic imaging

An intraoperative use of PA imaging that targets the epidermal growth factor receptor (EGFR) was described by Tummers et al. [54]. EGFR a transmembrane receptor is highly expressed in pancreatic cancer and hence can be used as a target for fluorescence imaging. Patients with pancreatic cancer, who were eligible for resection, were given a pretreatment dose of unlabeled cetuximab, which saturated the EGFR receptors in normal tissue with high expression. This also helped differentiate between a cetuximab reaction and a cetuximab-IRDye800 reaction. Thereafter, patients underwent surgery. During surgery, the abdomen was first imaged using laparoscopic optical imaging system (NIR imaging), which was followed by resection of tumor. Following resection, all excised tissues underwent ex vivo photoacoustic imaging. Lastly, the samples were sent for histopathological correlation. Results showed that during laparoscopic NIR imaging, the pancreatic tumor and lymph nodes would be identified in all patients (6/6). Using fluorescence imaging, a clear contrast was seen between the tumor and surrounding tissue for pancreatic cancer and lymph node dissection. Ex vivo photoacoustic images showed an increased PA signal in the primary tumor tissue as compared with the surrounding pancreatic tissue and in tumor-bearing LN. A difference in signal-to-noise ratio was also observed. Both these findings were consistent with the fluorescence imaging results in the same patients.

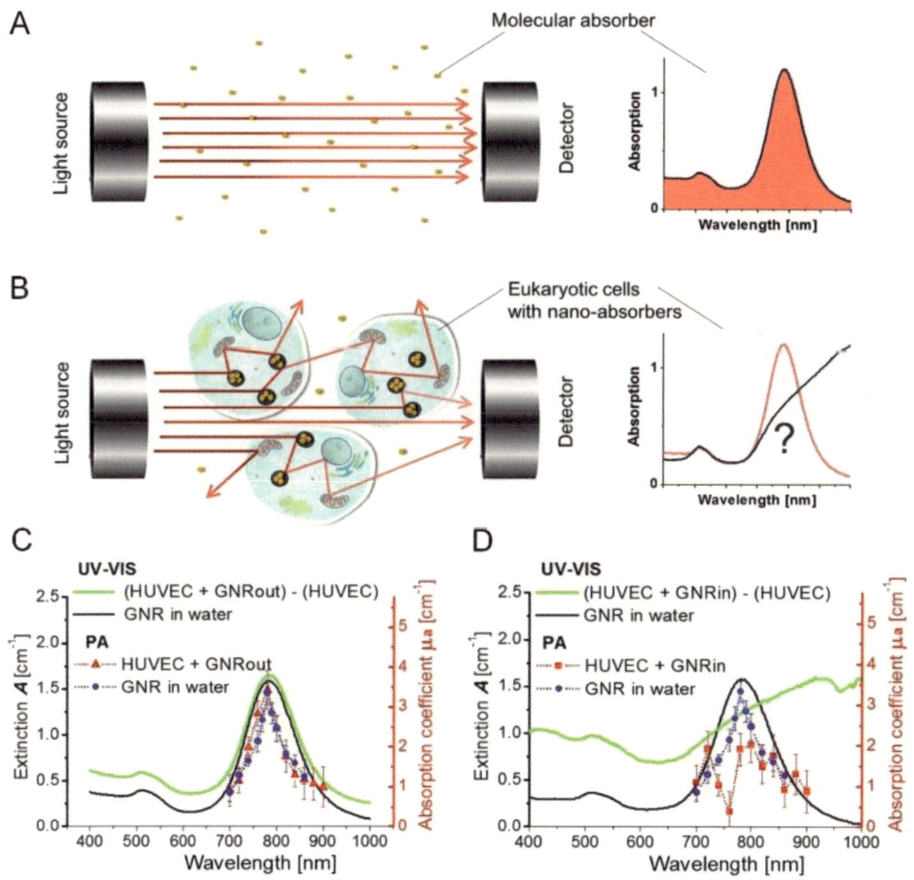

Figure 8.
(A) Absorption spectrum detected with UV-VIS spectrophotometer when light is attenuated only via molecular absorption. (B) UV-VIS spectrum when light is attenuated (i) due to molecular absorption by contrast nanoagents after its cellular uptake and (ii) a part of the incident photon flux is scattered. (C and D) absorption spectra of PEG-coated gold nanoparticles localized in extracellular matrix (HUVEC+GNRout) and internalized by HUVEC cells (HUVEC+GNRin) as detected by all-optical, narrow-beam PA spectrophotometer. Reproduced from open-source publication [22].

2.3.3 Identifying photoacoustic properties of target nanoagents in complex media

The future work on the detection and treatment of pancreatic cancer with PA will include investigation of theranostic nanoagents that have potential to provide specific molecular-targeted PA imaging of malignant tumor, PA and US activation of nanoagents, and following treatment of lesions.

Photoacoustic properties include (a) thermoelastic behavior of a system of an absorbing compound in complex surrounding environment and (b) absorption spectrum of nanoagents involved in biological processes. Petrova et al. [55] demonstrated that generation of PA response from blood in terms of its thermoelastic properties depends on compartmentalization of hemoglobin within erythrocyte. Recently, Pelivanov et al. [22] showed that localization of nanoagents in extracellular and/or intracellular matrix can affect the resultant absorption spectrum of nanoagents (**Figure 8**).

To identify thermoelastic and optical properties of PA contrast agents meticulously is crucial for successful implementation of theranostic nanoparticles for the therapy and early diagnostics of pancreatic cancer.

3. Conclusions

In summary, PA imaging is an emerging modality in gastroenterology and mostly used in combination with ultrasound or fluorescent imaging. Diagnosis through angiogenesis and blood oxygenation in the lining of esophagus and colon can be performed using PA imaging. High-resolution, multispectral PA imaging is offered via catheter-based (i) all-optical/noncontact solution and (ii) a miniaturized contact PA/US endoscope for preclinical studies. The capability to image both the mucosa and submucosa layers (up to 3–4 mm in depth of GI tract walls) is the vast advantage of these PA methods comparing to other routinely used optical imaging modalities. However, further miniaturization of catheter-based endoscopes and improvement on specificity of diagnostics through blood oxygenation parameters are required. Multispectral, multiscale PA/US system based on hand-held probe demonstrated rigorous data on assessment of a chronic inflammatory bowel disease in clinical studies. Using multispectral transabdominal examination, the PA system was able to evaluate hemoglobin level and blood oxygenation and accurately correlate it with abnormal activities in the large bowel and small intestine assessed through standard colonoscopic procedure. The only limitation of this method is depth beyond 5–6 cm in the body. Nonetheless, the proposed PA approach is noninvasive and use non-ionized radiation that is extremely fascinating for further clinical research. Many other PA imaging methods including intraoperative multimodality of PA and fluorescent imaging are currently under development and demonstrate high potential for use in gastroenterology.

Acknowledgements

This work was supported in part by the National Institutes of Health (National Cancer Institute) grant R01 CA181275.

Conflict of interest

The authors declare that there are no conflicts of interest on the presented manuscript.

Author details

Sheena Bhushan, Sharmila Anandasabapathy and Elena Petrova*
Baylor College of Medicine, Houston, Texas, USA

*Address all correspondence to: elena.petrova@bcm.edu

References

[1] Loukas M, Tubbs RS, Benninger B. Gastrointestinal Tract. Gray's Clinical Photographic Dissector of the Human Body. 2nd ed. Elsevier Health Sciences; 2018. pp. 198-222. DOI: 10.1016/b978-1-4377-2417-2.00015-5

[2] EBJ CV, Mouen K, Raman MV. In: Chandrasekhara V, editor. Clinical Gastrointestinal Endoscopy. 3rd ed. Philadelphia, Pennsylvania, USA: Elsevier; 2019

[3] Knyrim K, Seidlitz H, Vakil N, Classen M. Perspectives in "electronic endoscopy". Past, present and future of fibers and CCDs in medical endoscopes. Endoscopy. 1990;**22**(Suppl 1):2. DOI: 10.1055/s-2007-1012877

[4] Bruno MJ. Magnification endoscopy, high resolution endoscopy, and chromoscopy; towards a better optical diagnosis. Gut. 2003;**52**(Suppl 4): iv7-iv11. DOI: 10.1136/gut.52.suppl_4.iv7

[5] Wong Kee Song LM, Adler DG, Chand B, Conway JD, Croffie JMB, DiSario JA, et al. Chromoendoscopy. Gastrointestinal Endoscopy. 2007;**66**(4):639-649. DOI: 10.1016/j.gie.2007.05.029

[6] McGill SK, Evangelou E, Ioannidis JPA, Soetikno RM, Kaltenbach T. Narrow band imaging to differentiate neoplastic and non-neoplastic colorectal polyps in real time: A meta-analysis of diagnostic operating characteristics. Gut. 2013;**62**(12):1704-1713. DOI: 10.1136/gutjnl-2012-303965

[7] Tan NC-W, Herd MK, Brennan PA, Puxeddu RJBJO, Surgery M. The role of narrow band imaging in early detection of head and neck cancer. British Journal of Oral and Maxillofacial Surgery. 2012;**50**(2):132-136. DOI: 10.1016/j.bjoms.2010.12.001

[8] Kiesslich R, Goetz M, Neurath MF. Virtual histology. Best Practice & Research Clinical Gastroenterology. 2008;**22**(5):883-897. DOI: 10.1016/j.bpg.2008.05.003

[9] Goetz M, Watson A, Kiesslich R. Confocal laser endomicroscopy in gastrointestinal diseases. Journal of Biophotonics. 2011;**4**(7-8):498-508. DOI: 10.1002/jbio.201100022

[10] Giovannini M, Caillol F, Monges G, Poizat F, Lemaistre A-I, Pujol B, et al. Endoscopic ultrasound-guided needle-based confocal laser endomicroscopy in solid pancreatic masses. Endoscopy. 2016;**48**(10):892-898. DOI: 10.1055/s-0042-112573

[11] Wallace MB, Meining A, Canto MI, Fockens P, Miehlke S, Roesch T, et al. The safety of intravenous fluorescein for confocal laser endomicroscopy in the gastrointestinal tract. Alimentary Pharmacology & Therapeutics. 2010;**31**(5):548-552. DOI: 10.1111/j.1365-2036.2009.04207.x

[12] Izatt JA, Kulkarni MD, Hsing-Wen W, Kobayashi K, Sivak MV. Optical coherence tomography and microscopy in gastrointestinal tissues. IEEE Journal of Selected Topics in Quantum Electronics. 1996;**2**(4):1017-1028. DOI: 10.1109/2944.577331

[13] Yelbuz TM, Choma MA, Thrane L, Kirby ML, Izatt JA. Optical coherence tomography: A new high-resolution imaging technology to study cardiac development in chick embryos. Circulation. 2002;**106**(22):2771-2774. DOI: 10.1161/01.cir.0000042672.51054.7b

[14] Hatta W, Uno K, Koike T, Yokosawa S, Iijima K, Imatani A, et al. Optical coherence tomography for the staging of tumor infiltration in superficial esophageal squamous cell

carcinoma. Gastrointestinal Endoscopy. 2010;**71**(6):899-906. DOI: 10.1016/j.gie.2009.11.052

[15] Rösch T. Endoscopic ultrasonography: Imaging and beyond. Gut. 2003;**52**(8):1220-1226. DOI: 10.1136/gut.52.8.1220

[16] Cho CM. Training in endoscopy: Endoscopic ultrasound. Clinical Endoscopy. 2017;**50**(4):340-344. DOI: 10.5946/ce.2017.067

[17] Gusev VE, Karabutov AA. Laser Optoacoustics. New York: American Institute of Physics; 1993. p. 136. DOI: 0.1021/acsnano.7b01032

[18] Wang LV, SJs H. Photoacoustic tomography: In vivo imaging from organelles to organs. Science. 2012;**335**(6075):1458-1462. DOI: 10.1126/science.1216210

[19] Petrova E, Liopo A, Oraevsky AA, Ermilov SA. Temperature-dependent optoacoustic response and transient through zero Grüneisen parameter in optically contrasted media. Photoacoustics. 2017;7:36-46. DOI: 10.1016/j.pacs.2017.06.002

[20] Petrova EV, Brecht HP, Motamedi M, Oraevsky AA, Ermilov SA. In vivo optoacoustic temperature imaging for image-guided cryotherapy of prostate cancer. Physics in Medicine & Biology. 2018;**63**(6):064002. DOI: 10.1088/1361-6560/aab241

[21] Oraevsky AA, Karabutov AA. Optoacoustic Tomography. Boca Raton, London, New York, Washington, DC: CRC Press; 2003. DOI: 10.1117/12.910975

[22] Pelivanov I, Petrova E, Yoon SJ, Qian Z, Guye K, O'Donnell M. Molecular fingerprinting of nanoparticles in complex media with non-contact photoacoustics: Beyond the light scattering limit. Scientific Reports. 2018;**8**(1):14425. DOI: 10.1038/s41598-018-32580-2

[23] Pelivanov I, Buma T, Xia J, Wei C-W, O'Donnell M. A new fiber-optic non-contact compact laser-ultrasound scanner for fast non-destructive testing and evaluation of aircraft composites. Journal of Applied Physics. 2014;**115**(11):113105. DOI: 10.1063/1.4868463

[24] Wang L, Maslov K, Wang LV. Single-cell label-free photoacoustic flowoxigraphy in vivo. Proceedings of the National Academy of Sciences of the United States of America. 2013;**110**(15):5759-5764. DOI: 10.1073/pnas.1215578110

[25] Runge TM, Abrams JA, Shaheen NJ. Epidemiology of Barrett's esophagus and esophageal adenocarcinoma. Gastroenterology clinics of North America. 2015;**44**(2):203-231. DOI: 10.1016/j.gtc.2015.02.001

[26] Rastogi A, Puli S, El-Serag HB, Bansal A, Wani S, Sharma P. Incidence of esophageal adenocarcinoma in patients with Barrett's esophagus and high-grade dysplasia: A meta-analysis. Gastrointestinal Endoscopy. 2008;**67**(3):394-398. DOI: 10.1016/j.gie.2007.07.019

[27] Mansour NM, Groth SS, Anandasabapathy S. Esophageal adenocarcinoma: Screening, surveillance, and management. Annual Review of Medicine. 2017;**68**(1):213-227. DOI: 10.1146/annurev-med-050715-104218

[28] Muthusamy VR, Kim S, Wallace MB. Advanced imaging in Barrett's esophagus. Gastroenterology Clinics of North America. 2015;**44**(2):439-458. DOI: 10.1016/j.gtc.2015.02.012

[29] Ansari R, Zhang EZ, Desjardins AE, Beard PC. All-optical forward-viewing photoacoustic probe for high-resolution

3D endoscopy. Light: Science & Applications. 2018;7(1):75. DOI: 10.1038/s41377-018-0070-5

[30] Yang JM, Favazza C, Yao J, Chen R, Zhou Q, Shung KK, et al. Three-dimensional photoacoustic endoscopic imaging of the rabbit esophagus. PLoS One. 2015;**10**(4):e0120269. DOI: 10.1371/journal.pone.0120269

[31] Lim L, Streutker CJ, Marcon N, Cirocco M, Lao A, Iakovlev VV, et al. A feasibility study of photoacoustic imaging of ex vivo endoscopic mucosal resection tissues from Barrett's esophagus patients. Endoscopy International Open. 2017;**5**(8):E775-E783. DOI: 10.1055/s-0043-111790

[32] Ng SC, Shi HY, Hamidi N, Underwood FE, Tang W, Benchimol EI, et al. Worldwide incidence and prevalence of inflammatory bowel disease in the 21st century: A systematic review of population-based studies. The Lancet. 2017;**390**(10114):2769-2778. DOI: 10.1016/S0140-6736(17)32448-0

[33] Bhutiani N, Grizzle WE, Galandiuk S, Otali D, Dryden GW, Egilmez NK, et al. Noninvasive imaging of colitis using multispectral optoacoustic tomography. Journal of Nuclear Medicine: Official Publication, Society of Nuclear Medicine. 2017;**58**(6):1009-1012. DOI: 10.2967/jnumed.116.184705

[34] Knieling F, Neufert C, Hartmann A, Claussen J, Urich A, Egger C, et al. Multispectral optoacoustic tomography for assessment of Crohn's disease activity. The New England Journal of Medicine. 2017;**376**(13):1292-1294. DOI: 10.1056/NEJMc1612455

[35] Waldner MJ, Knieling F, Egger C, Morscher S, Claussen J, Vetter M, et al. Multispectral optoacoustic tomography in Crohn's disease: Noninvasive imaging of disease activity. Gastroenterology. 2016;**151**(2):238-240. DOI: 10.1053/j.gastro.2016.05.047

[36] Bray F, Ferlay J, Soerjomataram I, Siegel RL, Torre LA, Jemal A. Global cancer statistics 2018: GLOBOCAN estimates of incidence and mortality worldwide for 36 cancers in 185 countries. CA: A Cancer Journal for Clinicians. 2018;**68**(6):394-424. DOI: 10.3322/caac.21492

[37] Hidalgo M, Cascinu S, Kleeff J, Labianca R, Löhr JM, Neoptolemos J, et al. Addressing the challenges of pancreatic cancer: Future directions for improving outcomes. Pancreatology. 2015;**15**(1):8-18. DOI: 10.1016/j.pan.2014.10.001

[38] Stathis A, Moore MJ. Advanced pancreatic carcinoma: Current treatment and future challenges. Nature Reviews Clinical Oncology. 2010;7:163. DOI: 10.1038/nrclinonc.2009.236

[39] Bassi C, Salvia R, Butturini G, Marcucci S, Barugola G, Falconi M. Value of regional lymphadenectomy in pancreatic cancer. HPB : The Official Journal of the International Hepato Pancreato Biliary Association. 2005;7(2):87-92. DOI: 10.1080/13651820510028855

[40] Wagner M, Redaelli C, Lietz M, Seiler CA, Friess H, Büchler MW. Curative resection is the single most important factor determining outcome in patients with pancreatic adenocarcinoma. BJS. 2004;**91**(5):586-594. DOI: 10.1002/bjs.4484

[41] Verbeke CS, Gladhaug IP. Resection margin involvement and tumour origin in pancreatic head cancer. The British Journal of Surgery. 2012;**99**(8):1036-1049. DOI: 10.1002/bjs.8734

[42] Bronstein YL, Loyer EM, Kaur H, Choi H, David C, DuBrow RA, et al. Detection of small pancreatic tumors with

multiphasic helical CT. American Journal of Roentgenology. 2004;**182**(3):619-623. DOI: 10.2214/ajr.182.3.1820619

[43] Vahrmeijer AL, Hutteman M, Van Der Vorst JR, Van De Velde CJ, Frangioni JV. Image-guided cancer surgery using near-infrared fluorescence. Nature Reviews Clinical Oncology. 2013;**10**(9):507. DOI: 10.1038/nrclinonc.2013.123

[44] Kruskal J, Kane R. Intraoperative ultrasonography of the pancreas: Techniques and clinical applications. Surgical Technology International. Bellingham, Washington, USA. 1997;**6**:49-57. DOI: 10.1148/rg.307105051

[45] Sun MR, Brennan DD, Kruskal JB, Kane RA. Intraoperative ultrasonography of the pancreas. Radiographics. 2010;**30**(7):1935-1953. DOI: 10.15557/JoU.2015.0005

[46] Hata S, Imamura H, Aoki T, Hashimoto T, Akahane M, Hasegawa K, et al. Value of visual inspection, bimanual palpation, and intraoperative ultrasonography during hepatic resection for liver metastases of colorectal carcinoma. World Journal of Surgery. 2011;**35**(12):2779-2787. DOI: 10.1097/01.sla.0000171307.37401.db

[47] Handgraaf HJ, Boonstra MC, Van Erkel AR, Bonsing BA, Putter H, Van De Velde CJ, et al. Current and future intraoperative imaging strategies to increase radical resection rates in pancreatic cancer surgery. Biomed Research International. 2014;**2014**:8. Article ID: 890230. DOI: 10.1155/2014/890230

[48] Nelson DW, Blanchard TH, Causey MW, Homann JF, Brown TA. Examining the accuracy and clinical usefulness of intraoperative frozen section analysis in the management of pancreatic lesions. American Journal of Surgery. 2013;**205**(5):613-617. DOI: 10.1016/j.amjsurg.2013.01.015

[49] Harris PL, Rumley TO, Lineaweaver WC. Pancreatic cancer: Unreliability of frozen section in diagnosis. Southern Medical Journal. 1985;**78**(9):1053-1056. DOI: 10.1097/00007611-198509000-00007

[50] Witz M, Shkolnik Z, Dinbar A. Intraoperative pancreatic biopsy—A diagnostic dilemma. Journal of surgical oncology. 1989;**42**(2):117-119. DOI: 10.1002/jso.2930420210

[51] Homan K, Shah J, Gomez S, Gensler H, Karpiouk A, Brannon-Peppas L, et al. Combined Ultrasound and Photoacoustic Imaging of Pancreatic Cancer Using Nanocage Contrast Agents. Proceeding SPIE, Photons Plus Ultrasound: Imaging and Sensing. 2009; 71771M, 6 p. DOI: 10.1117/12.807520

[52] Homan KA, Souza M, Truby R, Luke GP, Green C, Vreeland E, et al. Silver nanoplate contrast agents for in vivo molecular photoacoustic imaging. ACS Nano. 2012;**6**(1):641-650. DOI: 10.1021/nn204100n

[53] Dai X, Qian W, Yang H, Yang L, Jiang H. Targeted molecular imaging of pancreatic cancer with a miniature endoscope. Applied Sciences. 2017;**7**(12):1241. DOI: 10.3390/app7121241

[54] Tummers WS, Miller SE, Teraphongphom NT, Gomez A, Steinberg I, Huland DM, et al. Intraoperative pancreatic cancer detection using tumor-specific multimodality molecular imaging. Annals of Surgical Oncology. 2018;**25**(7):1880-1888. DOI: 10.1245/s10434-018-6453-2

[55] Petrova EV, Oraevsky AA, Ermilov SA. Red blood cell as a universal optoacoustic sensor for non-invasive temperature monitoring. Applied Physics Letters. 2014;**105**(9):094103. DOI: 10.1063/1.4894635

Chapter 6

Photoacoustic Imaging of the Eye

Yanxiu Li and Yannis Mantas Paulus

Abstract

Photoacoustic imaging (PAI) is a novel, hybrid, non-ionizing, and non-invasive imaging technology with high-resolution, high sensitivity, high-contrast, and high depth of penetration. Hence, it has particularly useful applications in eye investigations. It can provide both anatomic and functional ocular characterizations. Many eye diseases, including macular degeneration and diabetic retinopathy, involve abnormalities in the vasculature, and thus the ability of PAI to affectively visualize the vasculature can be incredibly helpful to evaluate normal and disease states of the eye. In future research, PAI of the eye can be dramatically improved in terms of its resolution, use of contrast agents for molecular imaging, safety evaluations to develop a clinically approved system, and integration with existing fundus imaging modalities. Multimodality ocular imaging platforms have also been successfully developed by a combination of photoacoustic microscopy (PAM) with other optical imaging such as optical coherence tomography (OCT), scanning laser ophthalmoscopy (SLO), and fluorescence microscopy (FM). The multimodal images can accurately be acquired from a single imaging system and co-registered on the same image plane, enabling improved evaluation of eye disease states. In this book chapter, the potential application of photoacoustic imaging of the eye in both research and clinical diagnosis are comprehensively discussed as a powerful medical screening technique for visualization of various ocular diseases.

Keywords: photoacoustic imaging, photoacoustic ophthalmoscopy, photoacoustic microscopy, multimodal imaging

1. Introduction

The eye is unique organ in that it is optically transparent to visible light in normal health and thus offers optical access to internal structures. The refractive system of the normal eye includes the cornea, the fluid-filled anterior chamber, crystalline lens, and gel-like vitreous. These optically transparent structures have low absorption and light scattering to visible light, which plays a pivotal role in the entry of light and the formation of visual pathways. In the eye, the major endogenous light-absorbing materials are hemoglobin and melanin. Hemoglobin is present in blood vessels in the iris, choroid, ciliary body, retina, conjunctiva, and pathologic neovascularization. Melanin is present inside the uvea (iris, choroid and ciliary body), retinal pigment epithelium (RPE) and pigmented tumors.

Photoacoustic imaging (PAI), which converts light energy into sound waves, has been demonstrated to image hemoglobin [1, 2] and melanin [3, 4]. Thus eyes are an ideal system for PAI. PAI can provide an ocular vascular image based on the inherent optical absorbance of hemoglobin. Many eye diseases, including corneal neovascularization, macular degeneration, diabetic retinopathy, sickle cell retinopathy, and retinal vein occlusions, involve abnormalities in the vasculature. Thus the unique ability of PAI to provide detailed vascular morphologic information can assist ophthalmic diagnosis.

Currently several imaging modalities have been used to diagnose and monitor ocular diseases. Well-established imaging instruments include color fundus photography, ultrasound (US) imaging, fluorescein angiography (FA), indocyanine green angiography (ICGA), scanning laser ophthalmoscopy (SLO), optical coherence tomography (OCT) and OCT angiography (OCTA). Each imaging modality has advantages and disadvantages. OCT has a limited depth of penetration and has difficulty visualizing deep structures like the choroid and sclera. FA and ICGA require the use of an exogenous contrast agent that may cause adverse reactions and cannot specify the depth of the vessels. OCTA is unable to show leakage, provides limited view of microaneurysms, has a limited depth of penetration, and has a restricted field of view often with motion artifacts.

PAI is a new, emerging, non-ionizing and non-invasive imaging technology. PAI is based on the optical absorption contrast so that can provide anatomical information and functional analysis for the eye as well as with a high depth of penetration. In 2010, photoacoustic imaging was first described in the eye of a living animal [4]. Since then, there is a huge advance in photoacoustic imaging of the eye. Furthermore, the application of multimodality imaging systems including integrated photoacoustic microscopy (PAM) with other imaging modalities and the use of contrast agents for molecular imaging have undergone significant development. This chapter focuses on recent advances and applications of photoacoustic imaging of the eyes.

2. Photoacoustic imaging of the eye

Photoacoustic imaging systems can be divided into several categories which include: optical-resolution photoacoustic microscopy (OR-PAM), acoustic-resolution photoacoustic microscopy (AR-PAM), photoacoustic computed tomography (PACT), photoacoustic tomography (PAT) [5, 6]. Ocular PAI systems are primarily OR-PAM and AR-PAM systems due to the importance of high resolution imaging in the eye [7–9]. AR-PAM is used to image deep-tissues, where an illumination laser is diffusively delivered to the tissue and a focused acoustic detector is used to detect the induced PA signals. The lateral resolution of AR-PAM is determined by the acoustic focus spot, and the axial resolution is determined by the acoustic center frequency and bandwidth [5, 8]. OR-PAM has a tighter optical focus than AR-PAM and can achieve micrometer-level lateral resolution. The lateral resolution of OR-PAM is determined by the optical focal spot, and the axial resolution is still determined by the acoustic parameters [5, 8].

PAI can also be classified into mechanical-scanning systems and optical-scanning imaging systems [7]. In the mechanical-scanning mode, both the optical excitation and ultrasound transducer are simultaneously scanned over a planar surface. Each step of the scan can detect and generate a PA signal. During imaging, a water tank is necessary to maintain ultrasound coupling [6]. The mechanical-scanning can be used to visualize almost all the eye tissues, including the cornea, iris, lens, retina, choroid, sclera, and blood vessels [4]. In the optical-scanning mode, the focused optical illumination is raster-scanned using galvanometers, while the ultrasound transducer is kept stationary [10]. **Figure 1a** shows the schematic diagram of the optical-scanning in a multimodality system. **Figure 1b** illustrates the physical setup [11]. Compared with mechanical-scanning, optical-scanning can provide higher scanning speed, and it is suitable for chorioretinal microvasculature imaging, and compatibility with OCT and SLO [9, 12].

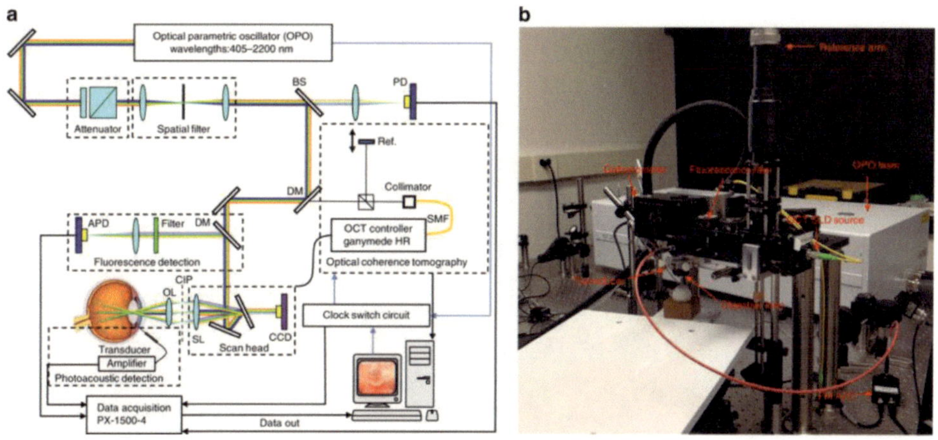

Figure 1.
Integrated multimodal PAM, OCT, and FM imaging system. (a) Schematic. (b) Photograph of the multimodality retinal and choroidal imaging system. OPO, optical parametric oscillator; SLD, superluminescent diode; BS, beam splitter; DM, dichroic mirror; SL, scan lens; OL, ophthalmic lens; CIP, conjugate image planes. Adapted with permission from Ref. [11] [Publisher: Springer Nature].

Currently, many applications of ocular PAI have been reported. This chapter summarizes those applications by the anatomic portion of the eye imaged and functional information.

2.1 PA imaging of the anterior segment vasculature of the eye

2.1.1 Normal iris vasculature

Anterior segment PA imaging of a normal eye receives signal from hemoglobin in the red blood cells of the iris microvasculature. Both AR-PAM and OR-PAM have been reported for iris vascular imaging. de la Zerda et al. [4] have used AR-PAM to qualitatively assess the rabbit eye *in vivo*. PA signals from the iris were detected, but the image of iris microvasculature was not clearly as the lateral resolution of the system was limited (around 200 μm). Hu et al. [13] have used a OR-PAM mechanical scanning system to get high resolution *in vivo* images of the iris microvascular in Swiss Webster albino mice (**Figure 2a**). The lateral resolution of the system was ~5 μm. The center frequency of the focused ultrasonic detector was 75 MHz. Hu et al. also have used two excitation wavelengths (570 and 578 nm) to estimate the concentrations of oxyhemoglobin (HbO_2) and deoxyhemoglobin (HbR), thereby have quantitatively measured sO_2 of the iris microvasculature. Dashed box in (**Figure 2a**) show the false-color map of the vasculature. With the limitation of mechanical scanning, the image acquisition took ~2 hours. Rao et al. [15] have used an OR-PAM hybrid-scanning system to image the major artery circle and the minor artery circle of the iris microvasculature in albino mice. Zhao et al. [14] have used a custom-built OR-PAM system to perform three-dimensional (3-D) vasculature imaging of rat iris *in vivo* with a lateral resolution of ~5 μm. They have extracted accurately vessel diameter, vascular density, and vascular tortuosity information (**Figure 2b**).

2.1.2 Corneal neovascularization

Normal corneal tissue is avascular and optically transparent. Corneal neovascularization is a pathologic condition that can lead to reduced vision characterized by the

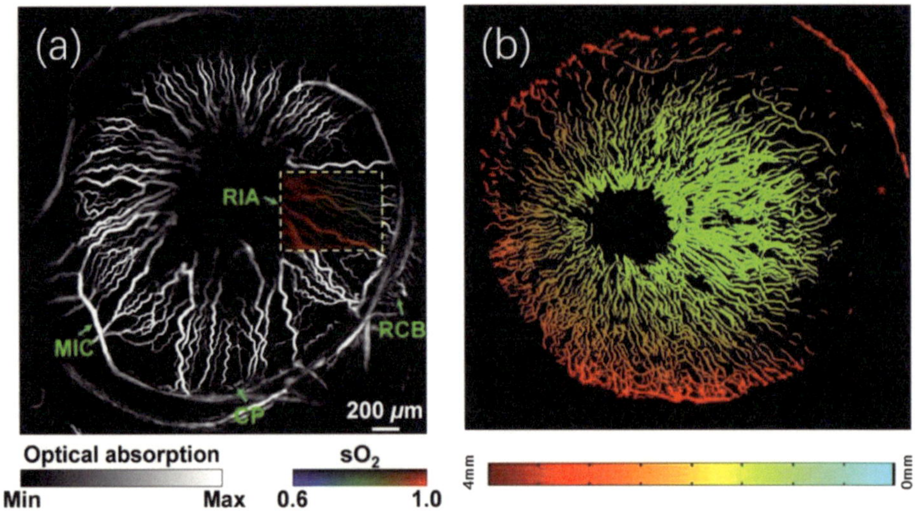

Figure 2.
PA images of the iris vasculature. (a) Label-free photoacoustic ophthalmic angiography of hemoglobin oxygen saturation sO_2 in the iris microvasculature of a living adult Swiss Webster mouse. CP, ciliary process; MIC, major iris circle; RCB, recurrent choroidal branch; RIA, radial iris artery. Adapted with permission from Ref. [13] [Publisher: The Optical Society (OSA)]. (b) Depth encoded photoacoustic imaging results of iris vasculature. Reproduced with permission from [14] [Publisher: SPIE].

presence of blood vessels in the cornea. This pathophysiology has been observed in humans with inflammatory or autoimmune responses.

Liu et al. [16] have used a mechanical scanning OR-PAM system to image corneal neovascularization of alkali-burn-injured mouse eyes *in vivo* for the first time. The lateral resolution of the system is 2.76 μm, and the axial resolution is around 50 μm. OR-PAM can achieve a similar lateral resolution as confocal microscopy without a contrast agent. However, the imaging depth is limited (700 μm) and the depth of variation of corneal vessels can be more than 1.4 mm owing to the curvature of the mouse cornea. Jeon et al. [17] have demonstrated a new method to evaluate blood circulation in the eye by combining *in vivo* PAM imaging. By using a machine learning algorithm, random sample consensus algorithm (RANSAC), they could have visualize the PA anterior ocular vasculature image and demonstrate layer-by-layer analysis of corneal neovascularization on a mouse eye (**Figure 3c–e**). This OR-PAM system had a 3 μm lateral resolution. PAM provides the ability to delineate corneal neovascularization from iris vessels and enables visualization of individual capillaries with high-resolution (**Figure 3b**). By stimulating corneal angiogenesis with an alkali burn in Tie2-GFP fluorescent-reporter mice, Kelly-Goss et al. [18] have used the PAM system to analyze functional hemodynamic changes in microvessels over time during adult corneal angiogenesis. These functional measurements of the microvasculature supplement traditional metrics of vascular architecture, including blood flow velocity, hemoglobin content, and sO_2. The study is the first time to measure the functional information on corneal neovascularization.

Several of the above studies of the anterior segment vasculature images have utilized mechanical scanning OR-PAM system. However, mechanical scanning OR-PAM has limitations: the imaging speed is slow and careful coupling medium is required. Recently, a non-interferometric photoacoustic remote sensing microscopy system has been developed with a high scan rate and an improved signal-to-noise (SNR).

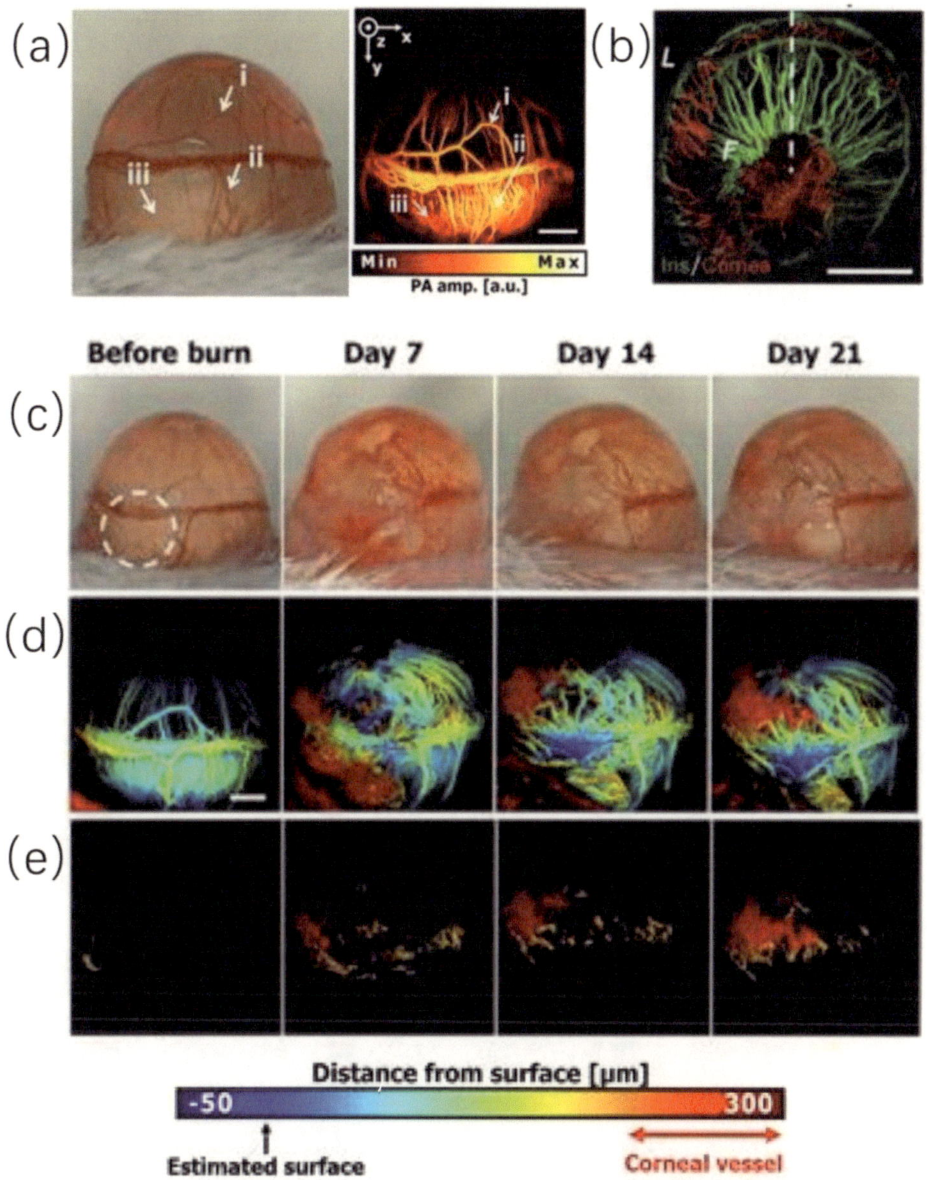

Figure 3.
PA images of the corneal neovascularization. (a) Photograph (left) and PA maximum intensity projection (MIP) image (right) of a mouse eye. Adapted with permission from Ref. [17] [Publisher: Springer Nature]. (b) PAM distinguishes between the iris (green) and corneal (red) vasculatures, due to their different depths. Adapted with permission from Ref. [18] [Publisher: Springer Nature]. (c–e) PA imaging of corneal neovascularization on a mouse alkali burn model eye. (c) Representative photographic images taken before alkali burn and 7, 14, and 21 days after alkali burn, the region of burn is highlighted with a white dashed circle. (d) Surface-based depth-encoded images. (e) Images after supra surface vessel isolation from the surface-based depth-encoded images. Scale bars indicate (a, c–e) 500 μm, (b) 50 μm. Adapted with permission from Ref. [17] [Publisher: Springer Nature].

2.2 PA imaging of the posterior segment vasculature of the eye

2.2.1 Retinal vasculature

Posterior segment PAI of a normal eye images red blood cells in the retinal and choroidal microvasculature. To obtain the images, both AR-PAM and OR-PAM systems have been used, along with mechanical scanning and optical scanning systems.

Using an AR-PAM system, de la Zerda et al. [4] have used a pulsed laser (wavelength 740 nm) combined with a 25 MHz central frequency transducer to mechanically scan a living rabbit eye. They have obtained the blood distribution of the posterior eye. But owing to the system's lateral resolution of 200 μm, limited features of the retinal or choroidal vasculature are visualized. Furthermore, it took 90 min to acquire the full 3-D image.

Jiao et al. [12] have developed an optical-scanning OR-PAM system for *in vivo* retinal imaging in 2010, which they have referred to as a photoacoustic ophthalmoscopy (PAOM). The pigmented rat fundus was imaging using the PAOM system (**Figure 4a–c**). In the PAOM system, the illumination source was a 532 nm pulse laser, the laser pulse energy was set lower than 40 nJ/pulse. The laser induced PA waves from the retina were detected by a custom-built needle ultrasonic transducer (30 MHz; bandwidth: 50%), which was placed in contact with the eyelid coupled by ultrasound gel. The lateral and axial resolution of the PAOM were ~20 and 23 μm, respectively. It only took 2.7 s for the volumetric PAOM images consisting of 256 B-scans. Multiple additional studies of imaging retinal vasculature by PAOM in small animals have been performed [10, 20–22]. PAOM as a non-invasive, 3-D microscopic imaging modality, has achieved high-speed, high-resolution *in vivo* imaging of the vasculature of the retina.

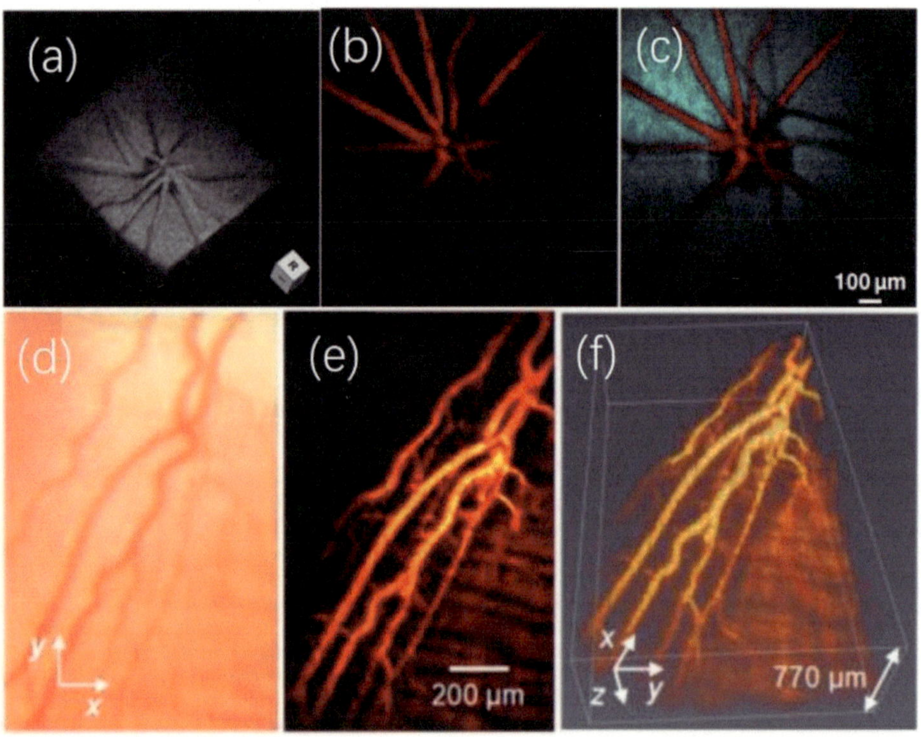

Figure 4.
Retinal vasculature imaging by optical-scanning OR-PAM system. (a–c) PA imaging of retinal vasculature in a rat eye. (a) Showing the imaged retinal structure by PAOM. (b) Segmented PAOM images of the retinal blood vessels. (c) Pseudo-colored PAOM images of the retinal vessels and RPE. Adapted with permission from Ref. [12] [Publisher: The Optical Society (OSA)]. (d–f) PA imaging of an albino rabbit retinal blood vessels. (d) Close-up of the retinal vessels of rabbit. (e) Maximum intensity projections of PAM signals of RVs and choroidal vessels. (f) 3D volumetric rendering of the PAM image. Adapted with permission from Ref. [19] [Publisher: The Optical Society (OSA)].

Before 2017, most reported PAM imaging of the posterior segment used small animals, like mice and rats. Mice have an axial length of ~3 mm and rats ~6 mm, which is much smaller that the human axial length of ~23 mm. In order to improve clinical translation of the technology, Tian et al. [19] have developed a novel PAM system to visualize rabbit chorioretinal vasculature. Rabbit eyes have an axial length of 18.1 mm [23], which is almost 80% of the axial length of human eyes. They have used a 570 nm pulsed laser. The excited PA signals have been captured by a custom-built needle-shaped ultrasonic transducer (27 MHz; bandwidth: 60%) placed in contact with the conjunctiva by ultrasound gel. The laser pulse energy on the rabbit cornea was 80 nJ or half the ANSI safety limit [24]. The lateral and axial resolution of the PAM system were 4.1 and 37 μm, respectively. This PAM system obtain high-resolution images of New Zealand rabbit retinal and choroidal vasculature (**Figure 4d–f**).

2.2.2 Choroidal vasculature

Wei et al. [25] have first applied an OCT-guided PAOM system to image the choroidal vessels in albino rats. The laser output optical wavelength was 578 nm, and the laser pulse energy was set below 40 nJ. The induced PA waves were detected by a custom-built, unfocused needle ultrasonic detector (35 MHz; bandwidth: 50%). The lateral resolution of PAOM was around 20 μm and the axial resolution of PAOM was 23 μm. Based on the OCT cross-sectional images, the distance from the retinal vessel layer to the choroid was found to be 200 μm [26]. Thus, the retinal and choroidal vessels were well resolved along the axial direction. **Figure 5a–c** show PAOM images of segmented retinal vessels and choroidal vessels. Song et al. [20, 27] have used an integrated PAOM and OCT system to detect *in vivo* posterior segment imaging of albino and pigmented rats. They have acquired both retinal and choroidal vessels from an albino rat by PAOM, but they were unable to visualize choroidal vessels from a pigmented rat due to melanin from the retinal pigment epithelium (RPE). **Figure 5d** and **e** show the contrast images. In addition, **Figure 5f** and **g** show the images of choroidal vessels in albino rabbit obtained by Tian et al. [19]. These experimental results have demonstrated that PAI can visualize the choroid and retinal vessels. However, PAM was limited to albino animals when imaging the choroid. The high melanin concentration of the RPE present in front of choroidal vasculature did not allow the illumination light with in visible spectrum range to penetrate through the RPE to generate sufficient PA signal from the choroid for imaging. Near infrared (NIR) light may be a solution to allow PAOM to image choroidal vessels in pigmented eyes.

2.2.3 Retinal neovascularization

Retinal neovascularization (RNV) is the growth of abnormal new retinal blood vessels and represents a major cause of vision loss and blindness. RNV is a common complication of several retinal diseases, including proliferative diabetic retinopathy (PDR), retinal vein occlusions (RVO), sickle cell retinopathy, and retinopathy of prematurity (ROP).

Zhang et al. [11] have demonstrated an integrated PAM, OCT, and fluorescence microscopy (FM) multimodality system to evaluate RNV *in vivo* in rabbit eyes. The model of RNV was induced by intravitreous injection of vascular endothelial growth factor (VEGF), which mimics numerous retinal diseases, such as PDR. Using a similar PAM system as Tian et al. in 2017 but with a laser wavelength of 532 nm, this study has presented high-quality visualization of retinal vasculature and RNV in

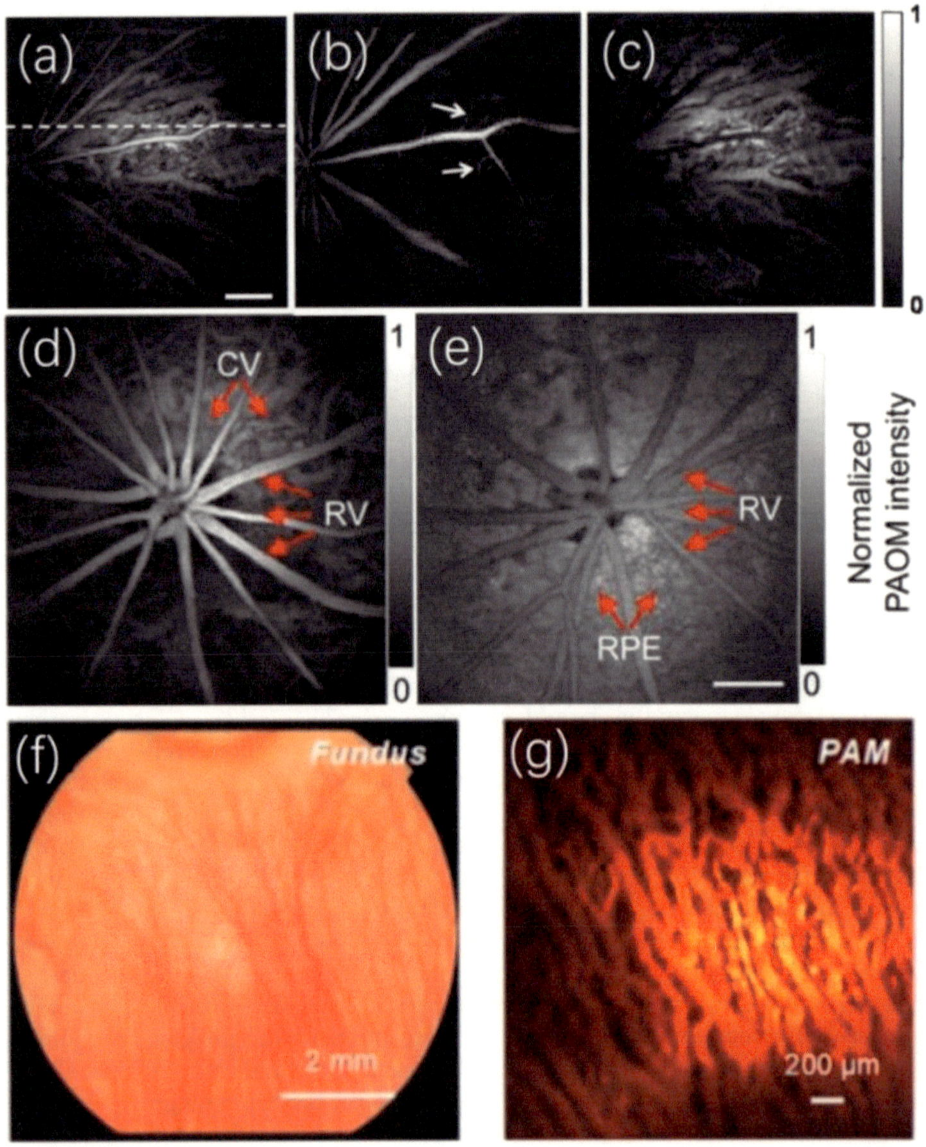

Figure 5.
PA images of the chorioretinal vasculature. (a–c) PA imaging of chorioretinal vessel in an albino rat eye. (a) Complete chorioretinal vessel network image. (b) Segmented retinal vessels. (c) Segmented choroidal vessels [25]. (d, e) Comparing PA images of the chorioretinal vasculature in an albino rat and a pigmented rat. (d) Acquired from an albino rat, and (e) acquired from a pigmented rat. RV: retinal vessel; CV: choroidal vessel; RPE, retinal pigment epithelium. Scale bars indicate 500 µm adapted with permission from Ref. [27] [Publisher: JoVE]. (f, g) PA imaging of an albino rabbit. (f) Fundus photograph of rabbit demonstrating the choroidal vessel distribution inferior to the optic nerve. (g) Maximum intensity projections of PAM signals of the CVs. Adapted with permission from Ref. [19] [Publisher: The Optical Society (OSA)].

albino and pigmented rabbits *in vivo*. **Figure 6** shows that the blood vessel diameter and blood flow in neovascularization are smaller than those of normal vessels, yet PAM can detect neovascularization and the small, irregular vascular network. At the same time, Nguyen et al. [28] have found that multi-wavelength PAM can selectively monitor and detect RVO and RNV in the rabbit retina. This has demonstrated that PAM could achieve label free imaging of microvasculature and RNV without administration of exogenous contrast agents. In addition, rabbit eyes have a similar size to human eyes, which paves the road to clinical translation of the technology.

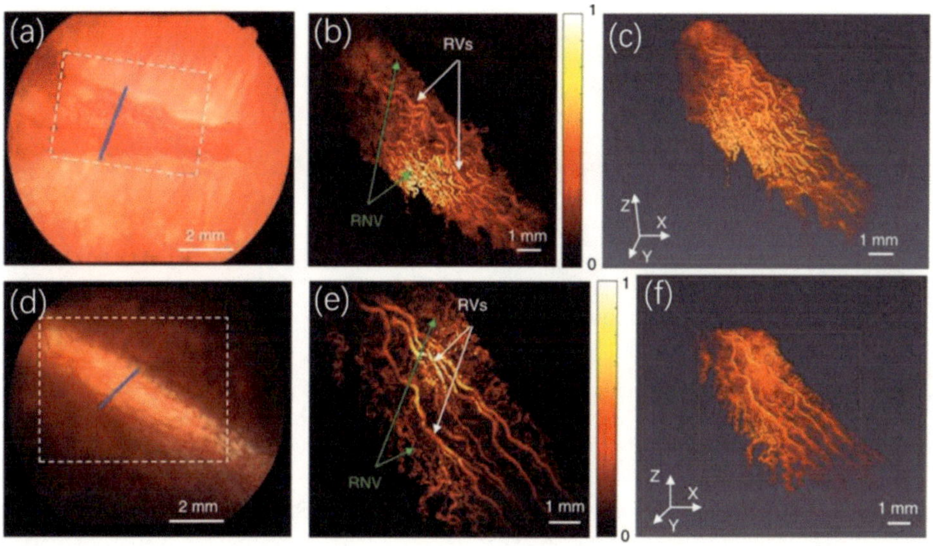

Figure 6.
PA images of the retinal neovascularization. (a–c) RNV in albino rabbits. (a) Color fundus image. (b) 2D PAM image indicated by white dashed box in (a). (c) 3D reconstruction of PAM image. (d–f) RNV in pigmented rabbits. (d) Color fundus image. (e) 2D PAM image indicated by white dashed box in (d). (f) 3D reconstruction of PAM image. White arrows indicate normal retinal vessels; green arrows indicate RNV. Adapted with permission from Ref. [11] [Publisher: Springer Nature].

2.2.4 Choroidal neovascularization

Choroidal neovascularization (CNV) is characterized by the abnormal growth of new vessels originating from the choroidal vasculature and their subsequent growth under the RPE, subretinal space, or a combination of both [29]. CNV most commonly occurs in neovascular age-related macular degeneration (AMD), which is a major cause of vision loss. Dai et al. have used PAM system to investigate laser-induced rat CNV evolution. For the PAM system, the laser output wavelength was 532 nm, and the pulse energy was 60 nJ. The induced PA signals from the retina were detected by a customized ultrasonic transducer (30 MHz; bandwidth: 15 MHz). the axial and lateral resolution of PAM were 50 and 20 μm, respectively. The new capillaries growing from the choroid up to subretinal space can be distinguished as shown in **Figure 7**.

2.3 PA imaging in retinal metabolism

Retinal oxygen metabolic rate ($rMRO_2$) is an essential parameter in the retina and can help further understanding of some blinding diseases, such as diabetic retinopathy [31, 32] and glaucoma [33, 34]. The precise measurement of $rMRO_2$ can be critical in investigating these blinding diseases. Song et al. [35] and Liu et al. [36] have successfully determined $rMRO_2$ in rats by integrating PAOM and OCT. Obtaining $rMRO_2$ measurements required measuring retinal blood flow and sO_2 together. They have quantified retinal blood flow by Doppler SD-OCT, and retinal sO_2 by PAOM at three wavelengths (570, 578, and 588 nm). Owing to the distinct light absorption spectrum between HbO_2 and HbR, multi-wavelength imaging can assess the sO_2 in retinal vessels. They have calculated total retinal blood flow as 7.43 ± 0.51 and 7.38 ± 0.78 μL/min within the venous and arterial systems, respectively. The sO_2 value in arterial and venous blood were 93.0 ± 3.5 and 77.3 ± 9.1%, respectively. In PAOM system, they used a

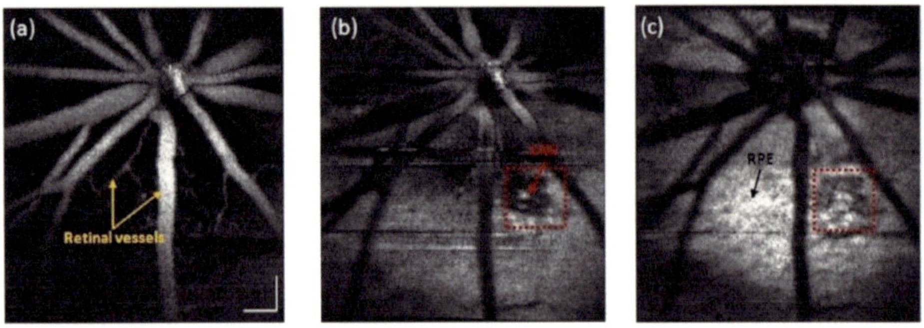

Figure 7.
PAM images at different depth in CNV rat fundus. (a) 2D enface image of anterior retinal structure. (b) 2D enface image of posterior retinal structure. (c) 2D enface image of RPE layer. Scale bars indicate 100 μm. Adapted with permission from Ref. [30] [Publisher: SPIE].

tunable laser (output laser wavelength: 570, 578, and 588 nm), the laser energy was 40 nJ/pause. Acoustic waves were detected by an unfocused small-footprint ultrasonic transducer (40 MHz; bandwidth: 30 MHz). The lateral resolution was 20 μm, and the axial resolution was 23 μm. The study measured $rMRO_2$ in normal small animals. Hariri et al. [37] have demonstrated the use of PA ocular imaging (PAOI) in measuring chorioretinal oxygen saturation ($CR\text{-}sO_2$) gradients in New Zealand white rabbits with ocular ischemia model. The PAOI signal showed a sixfold decrease in $CR\text{-}sO_2$ after significant elevation of IOP during ischemia. In the PAOI, they used a tunable laser (680–970 nm), 750 and 850 nm were used to differentiate the HbO_2 and HbR. A linear array ultrasound transducer with a center frequency of 15 MHz was used to detect acoustic signals. The lateral and axial resolution were 580 and 290 μm, respectively. One limitation of the study was balancing special resolution and penetration depth. Although they could achieve much deeper penetration, they could not discriminate between the retina and the choroid, nor could they distinguish individual vessels.

2.4 PA imaging of melanin of the eye

Melanin is naturally present in the eye within the iris, ciliary body, pigmented choroid, and RPE [38]. The RPE is a single layer of epithelial cells beneath the neurosensory retina and tightly adherent to the underlying choroid. The RPE plays a crucial role in the overall health of the retina: nourishing photoreceptors and disposing of retinal waste and metabolic end products. The melanin concentration in RPE can decrease over time due to light exposure and oxidative stress [39]. RPE melanin decrease is a sign of ocular senescence and is a risk factor and a pathognomonic sign of AMD [40, 41]. To further study melanin, PAI has been used to specifically detect and quantify melanin [42].

In 2010, Silverman et al. [43] for the first time have demonstrated that PAI can acquire images of melanin in the iris in *ex vivo* pig eye. Jiang et al. [44] have designed an adaptive optics PAM (AO-PAM) system to correct the wavefront errors of the illuminating light of PAM. The lateral resolution of PAM was improved to 2.5 μm with AO. They imaged the ciliary body and the RPE layer of an *ex vivo* pig eye. In addition, the single RPE cells were first resolved by AO-PAM system (**Figure 8a** and **b**). Jiao et al. [12] have successfully acquired images of the retinal vessels and RPE in pigmented rat eyes by using integrated PAOM with OCT system with a 23 μm axial resolution. The study was the first *in vivo* demonstration of PA imaging melanin of eye. Zhang

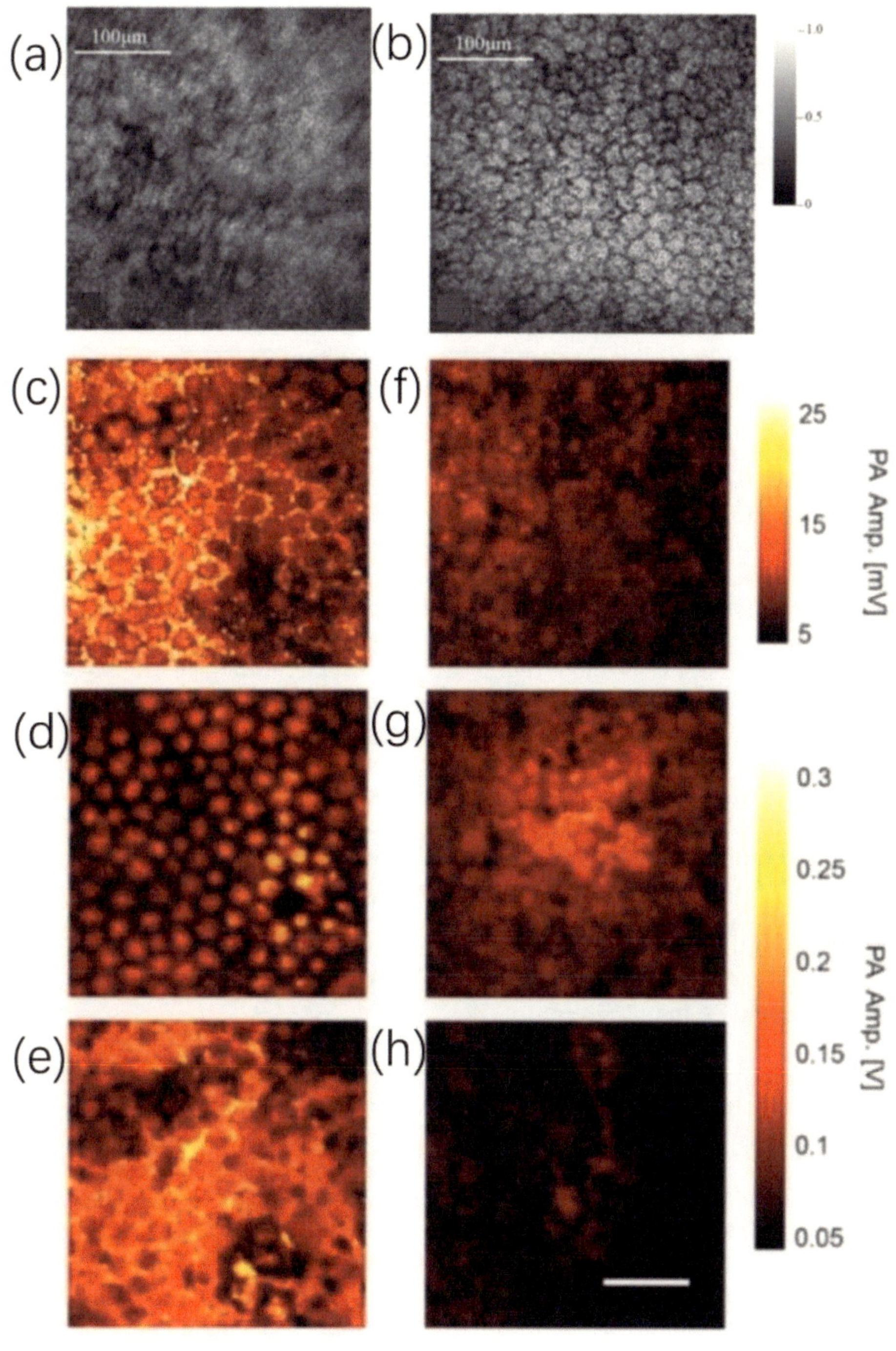

Figure 8.
PA images of melanin of the RPE. (a, b) MAP of the PAM images of the RPE of the pig eye: (a) without AO. (b) With AO. Adapted with permission from Ref. [44] [Publisher: The Optical Society (OSA)]. (c–h) PA images of porcine and human RCC. (c) PA image of porcine RCC acquired by customized piezoelectric transducer. (d, e) PA image of porcine RCC acquired by broadband MRR detector. (d) Axially segmented porcine RPE. (e) Axially segmented porcine choroid. (c, e) Are images of the same area on a single sample. (f) PA image of human RCC acquired by customized piezoelectric transducer. (g, h) PA image of human RCC acquired by homemade MRR detector. (g) Axially segmented human RPE. (h) Axially segmented human choroid. (f, h) Are images of the same area on a single sample. RCC, RPE-choroid complex. Bottom right scale bar indicates 50 μm. Adapted with permission from Ref. [45] [Publisher: The Optical Society (OSA)].

et al. [46] have combined autofluorescence (AF) with PAOM system to image the RPE melanin and lipofuscin in albino and pigmented rat eyes. Song et al. have developed a multimodal system that contains PAOM, SLO, OCT, and FA to image the eye. They have demonstrated that melanin from the RPE and choroid was able to be imaged *in vivo* in the retinal of albino and pigmented rats, [20] as well as mice [21]. To demonstrate the feasibility of NIR-light PAOM *in vivo*, Liu et al. [47] have developed a dual-wavelength (532 and 1064 nm) PAOM and further integrated it with SD-OCT to image albino and pigmented rat eyes. They found that NIR light (1064 nm) PAOM system successfully imaged RPE/choroidal melanin with good signal-to-noise ratio (SNR) in pigmented rat eyes. After that, in 2015, Liu et al. [48] have demonstrated the feasibility of an OCT-PAM system working in the NIR-light. The OCT and PAM used the same broadband light source with a center wavelength of 800 nm. This system can provide a deep-penetration depth and melanin-specific images of the retina in pigmented rats.

The above papers are qualitative studies of eye melanin by PAI. PAI also has the capability of providing a quantitative reading of melanin concentration in the eye. Shu et al. [49] have performed a Monte Carlo (MC) stimulation to investigate light propagation and energy deposition in the eye and tested the feasibility of using PAOM to quantify the RPE melanin concentration. In this study, they used PA signals from retinal blood vessels as references to measure RPE melanin variation. However, the main challenge in accurate quantification of RPE melanin concentration is to distinguish the RPE from underlying pigmented choroid. The RPE is a 10 μm thick cell monolayer and is tightly attached to the choroid, and thus a higher axial resolution of PAOM was necessary. Shu et al. [45] then have used a micro-ring resonator (MRR) detector in their PAOM system to increase the axial resolution (<10 μm). They obtained images where the RPE and choroid can be distinguished in *ex vivo* porcine and human eyes and quantified the absolute melanin concentrations in the RPE and pigmented choroid, respectively (**Figure 8c–h**).

PAI can obtain volumetric images of melanin in the eye and can quantify melanin in the RPE and the choroid. However, the technique has yet to be explored in eye disease models and the human eye, so it is unknown that how it can be used in eye research and clinic in the future.

2.5 Multimodal PA imaging system

PA imaging can obtain structural and functional imaging of the eye. PAI can also be combined with other ocular imaging devices, such as OCT, SLO, AF, AO, FA, and fundus photography. These combined multimodal platforms can compensate the weaknesses of each system and provide more comprehensive information of the eye. That would be very beneficial for investigating ocular pathology and detecting disease. For instance, with correcting wave-front errors with AO, the lateral resolution of PAM can be improved [44]. With PAM and AF combined, information on the distribution and concentration of retinal melanin and lipofuscin can be obtained *in vivo* simultaneously [46]. The most commonly integrated imaging systems combine PAI with OCT since this carries several advantages. OCT can not only acquire posterior segment anatomic information with high axial and lateral resolutions, but also can quantitatively measure retinal blood velocity and flow rate. In addition, OCT can also guide PAI, helping the PAI system image a specific region of interest in the posterior segment of the eye. Song et al. [20, 21] have developed a multimodality system which integrated PAOM, SD-OCT, AF-SLO, and FA to provide optical absorption, optical backscattering, and fluorescence properties of the retina in albino and pigmented rats and mice. Song and Liu et al. [35, 36] have quantified the $rMRO_2$ using PAOM and SD-OCT. PAOM measured the sO_2, and SD-OCT mapped the blood flow rate. Tian and Nguyen et al. [19, 28, 50, 51] have used a PAM and OCT dual-modality system to acquire chorioretinal imaging and retinal neovascularization imaging in living rabbit eyes. Zhang et al. [11] have described a multimodality imaging system that combines PAM, OCT, and fluorescence microscopy (FM) to evaluated retinal angiogenesis in albino and pigmented rabbit eyes. The multimodal images can be acquired from a single imaging system and co-registered on the same image plane. Thus, ocular diseases can precisely be detected and visualized at earlier stages, resulting in improved understanding of pathophysiology, improved management, and improved monitoring of retinal treatment to prevent vision loss.

3. Limitations and future directions

3.1 Innovative laser sources for PA imaging

PA imaging can achieve anatomic information with monochromatic pulsed laser illumination. However, to acquire ocular functional information using PAI, multi-wavelength laser illuminations is required. An optical parametric oscillator (OPO) (for example, the Ekspla NT-242, pulse repetition rate 1 kHz, duration 3–6 ns, tunable wavelength range 405–2600 nm) can be used as a suitable laser source to obtain the above information. However, in multimodality systems, such as PA imaging and OCT, the two modalities can require different types of illuminations and thus separate light sources. Liu et al. [48] have used a single ultrafast laser source (pulse repetition rate 10 kHz, pulse duration 3 ns, center wavelength 800 nm; bandwidth 30 nm) to acquire melanin-specific images of *in vivo* rat retina simultaneously from PAM and OCT. However, this shared laser was limited to imaging melanin since the absorption of hemoglobin was relatively weak in the NIR range. Recently, pulse laser diodes (PLD) have been described as an alternative laser excitation source in mouse ear vascular structures [52]. In the future, one could envision PLD and other novel laser sources for PAI and multimodal imaging applied to ophthalmology.

3.2 Non-contact detection of PA signal

In reported studies to date, researchers either placed a water tank on the cornea or directly coupled an ultrasonic needle transducer with the eyelid or sclera using ultrasonic gel or balanced salt solution (BBS). Given the sensitivity of the human eye, this is not ideal for clinical application on patients. Non-contact remote sensing has been described where an interferometer is used to sense the surface displacement induced by PA ultrasound signals [53, 54]. Discovering a non-contact ultrasonic detection with both stability and sensitivity is a promising area of active investigation.

3.3 Field of view of PA imaging

The field of view (FOV) of PA imaging is limited by the ultrasound transducer. The typical scan area of reported studies has been shown in **Table 1**. For most studies, the image area was no more than 3 × 3 mm. The human retina area is about 10 cm^2, so increasing the field of view would be beneficial, possibly through novel ultrasound transducer configurations or array patterns.

3.4 Speed of PA image acquisition

Several current PA imaging systems can acquire high-resolution images in <1 min. The imaging speed is limited by the laser pulse repetition rate. The image

PA system	Applications	Publication	Scanning pattern	Laser illumination	US	Speed and size	Resolution	Animal
AR-PAM	whole eye	de la Zerda [4]	Wide field illumination	W: 700/740 nm E: 0.5 mJ/cm^2	CF: 15/25 MHz B: 60%	90 min for 12 × 8 mm	L: 200/240 μm A: 83/50 μm	rabbit *in vivo*; pig *ex vivo*
OR-PAM	Iris vasculature and sO_2	Hu [13]	Mechanical scanning	W: 570/578 nm E: 40 nJ	CF:75 MHz	120 min for 2 × 2 mm	L: 5 μm A: 15 μm	Swiss Webster mouse
OR-PAM	Iris vasculature	Rao [15]	Hybrid-scanning	W: 532 nm E: 60 nJ	B: 25 MHz	128 s for 4 × 4 mm	L: 3.5 μm A: <31 μm	Albino mouse
OR-PAM	Iris vasculature	Zhao [14, 55]	Mechanical scanning	W: 532 nm E: 11 mJ/cm^2	CF: 75 MHz	0.5 frame/s	L: 5.43 μm	SD rat
OR-PAM	Corneal NV; iris vasculature	Liu [16]	Mechanical scanning	W: 532 nm E: 80 nJ	CF:10 MHz B: 80%	20 min for 3 × 3 mm	L: 2.76 μm A: 50 μm	C57BL6/J mouse
OR-PAM	Anterior vasculature	Jeon [17]	Mechanical scanning	W: 532 nm	CF: 50 MHz B: 50 MHz	3 × 3 mm	L: 3 μm	BALB/c mouse
OR-PAM	Corneal NV; blood flow and sO_2	Kelly-Goss [18]	Mechanical scanning	W: 532/559 nm E: 50/30 nJ	CF:35 MHz B: 70%	30 min for entire cornea	2.7 μm	Transgenic Tie2-GFP mouse
PAOM+SD-OCT	Retinal vasculature; RPE	Jiao [12]	Optical scanning	W: 532 nm E: 40 nJ	CF: 30 MHz B: 50%	2.7 s for 2 × 2 mm	L: 20 μm A: 23 μm	Long Evans rat
PAOM+SD-OCT	Retinal and choroidal vasculature	Wei [25]	Optical scanning	W: 578 nm E: 40 nJ	CF: 35 MHz B: 50%	2.7 s for 2 × 2 mm	L: 20 μm A: 23 μm	SD rat
PAOM+SD-OCT + AF-SLO + FA	Retinal and choroidal vasculature; RPE	Song [20]	Optical scanning	W: 532 nm E: 40 nJ	CF: 40 MHz B: 15 MHz	2.7 s for 2 × 2 mm	L: 20 μm A: 23 μm	Albino / pigmented rats
Fundus camera +PAOM	Retinal and choroidal vasculature	Liu [22]	Optical scanning	W: 532 nm E: 60 nJ	CF: 35 MHz B: 50%	2.7 s for 2 × 2 mm	L: 20 μm A: 23 μm	Albino / pigmented rats
PAOM+SD-OCT	Retinal and choroidal vasculature; RPE	Song [27]	Optical scanning	W: 532 nm E: 40 nJ	CF: 40 MHz B: 16 MHz	2.7 s for 2 × 2 mm	L: 20 μm A: 23 μm	Albino / pigmented rats

PA system	Applications	Publication	Scanning pattern	Laser illumination	US	Speed and size	Resolution	Animal
PAM+OCT	Retinal and choroidal vasculature	Tian [19]	Optical scanning	W: 570 nm E: 80 nJ	CF: 27 MHz B: 60%	65 s for 3 × 3 mm	L: 4.1 μm A: 37 μm	Albino rabbit
PAM+OCT + FM	Retinal vasculature and RNV	Zhang [11]	Optical scanning	W: 532 nm E: 80 nJ	CF: 27 MHz B: 60%	65 s for 3 × 3 mm	L: 4.1 μm A: 37 μm	Albino/ pigmented rabbits
PAM+OCT	RVO and RNV	Nguyen [28]	Optical scanning	Multi-wavelength E: <80 nJ	CF: 27 MHz B: 60%	65 s for 3 × 3 mm	L: 4.1 μm A: 37 μm	Albino rabbit
PAM+OCT	CNV; retinal structure; RPE	Dai [30]	Optical scanning	W: 532 nm E: 60 nJ	CF: 40 MHz B: 16 MHz	19.5 frame/s	L: 20 μm A: 50 μm	Brown Norway rat
PAOM+OCT	Retinal $rMRO_2$; sO_2; blood flow	Song [35], Liu [36]	Optical scanning	W:570/578/588 nm E: 40 nJ	CF:30 MHz B: 15 MHz	2.7 s for 2 × 2 mm	L: 20 μm A: 23 μm	SD rat
OR-PAM	Chorioretinal sO_2	Hariri [37]	Optical scanning	W: 750/850 nm E: 8 ± 0.5 mJ	CF: 15 MHz	0.8 frame/s	L: 580 μm A: 290 μm	Albino rabbit
AO-PAM	Ciliary body, RPE melanin	Jiang [44]	Optical scanning	W:532 nm E: 40 nJ	CF:30 MHz B:50%	2.7 s for 2 × 2 mm	L: 2.5 μm A: 23 μm	Pig *ex vivo*
PAOM+AF	RPE melanin; lipofuscin	Zhang [46]	Optical scanning	W: 532 nm E: 40 nJ	CF: 30 MHz B: 50%	2.7 s for 2 × 2 mm	L: 4.5 μm A: 23 μm	Albino/ pigmented rats
PAOM+SD-OCT	RPE melanin	Liu [47]	Optical scanning	W: 1064/532 nm E: 20 nJ	CF:30 MHz B: 50%	2.7 s for 2 × 2 mm	L: 20 μm A: 23 μm	SD/Long Evans rats
Broadband PAM	RPE melanin concentration	Shu [45]	Optical scanning	W: 532 nm	MRR B: 280/27 MHz	22 s per image	A: <10 μm	Porcine/human RCCs *ex vivo*

W, wavelength; E, energy; CF, central frequency; B, bandwidth; NV, neovascularization; SD, Sprague Dawley; RPE, retinal pigmented epithelial; RNV, retinal neovascularization; RVO, retinal vein occlusion; CNV, choroidal neovascularization; $rMRO_2$, retinal oxygen metabolic rate; AO, adaptive optics; AF, autofluorescence; MRR, micro-ring resonator; RCC, RPE-choroid complexes.

Table 1.
Summary of key articles demonstrating photoacoustic imaging of the eye.

speed can be reduced until it depends on the ultrasound propagation time from the posterior eye. The advantage of improving imaging speed is that it will reduce motion artifacts and increase patient comfort. Ideally a clinical PA system would acquire images in less than a second to a couple seconds.

3.5 Lateral and axial resolution of PA

The lateral resolution of PA imaging is determined by the smallest achievable spot size of the illuminating light in the retina. Tian and Zhang et al. described the lateral resolution of their PAM system as 4.1 μm [11, 19]. AO allows one to correct for ocular aberrations and has been applied to ophthalmic imaging to improve resolution. A feasibility study was conducted on the integration of AO with PAM for *ex vivo* imaging of the eye, which have improved the lateral resolution of the PAM to 2.5 μm. If one adds AO into multimodality systems, it can further improve the lateral resolution of PA imaging.

The axial resolution of PA imaging is determined by the bandwidth of the ultrasound transducer. The larger ultrasonic bandwidth will allow higher axial resolution, but with less sensitivity. Therefore, a balance is required between axial resolution and sensitivity.

3.6 Real-time PA imaging

Currently many PA images require post-image processing to display images. Further investigation of an advanced ophthalmic PA imaging system is to obtain real-time imaging.

3.7 Imaging the choroidal vasculature in pigmented animals

PA imaging is often limited to albino animals when imaging the choroid due to the high PA signal of the RPE melanin in pigmented animals. Further methods need to be developed and refined to penetrate the RPE layer and allow improved choroidal visualization. NIR light has been proposed as a solution, but further testing is needed.

3.8 Imaging ocular disease animal models

PA imaging has been demonstrated in some animal eye models to date, including mouse corneal neovascularization model, rat CNV model, and rabbit retinal neovascularization model. PA imaging needs to be further explored in more eye disease models to evaluate how the information provided by PA imaging will be used clinically in the future.

3.9 Exogenous contrast agents and theranostics

To improve PA image sensitivity and specificity, exogenous contrast agents can be utilized. PA imaging with exogenous contrast agents also allows one to extend the imaging scope to molecular imaging [56, 57]. Exogenous contrast agents for PA imaging include organic and inorganic agents. However, each agent has advantages and limitations. Organic agents (e.g., ICG) can have a limited level of contrast enhancement but a more rapid path to clinical translation [58]. Inorganic agents (e.g., gold nanoparticle (AuNP)) can offer higher contrast but a less rapid path to clinical translation due to less long-term evidence of biosafety [59, 60]. Thus, exploring suitable exogenous contrast agents with safety and high contrast can be used for PA imaging will be meaningful in future research. In addition, theranostic agents that can be used both for diagnostic imaging and therapy, should be further refined and developed to allow for targeted therapy at the time of imaging.

3.10 Safety evaluations and clinical approval

Before PA imaging can be applied to human imaging, a thorough evaluation of laser safety is necessary. Although the reported systems compare their laser fluence to the ANSI laser safety regulations, [24] one must monitor the long-term effect of both single and multiple imaging sessions on the structure and function of the retina. In additional, regulatory approval should be sought for a clinical system so that PA can be applied to patients and diseases.

4. Conclusion

The chapter introduces the applications, recent developments, and future directions of PA imaging in the eye. It has been demonstrated that PA imaging can provide both anatomic and functional information of eye with high-resolution, high sensitivity, high-contrast, and high depth of penetration. This chapter describes the ocular structure of PA imaging including normal vasculature of the iris, retina, and choroid, neovascularization in cornea, retina, and choroid, and melanin of the RPE. This chapter summarized PA imaging to quantify the functional information of measuring the vascular sO_2 and quantifying the absolute melanin concentrations. Limitations and future directions of PA imaging of the eye are also discussed.

Acknowledgements

This research was supported by a grant from the National Eye Institute 1K08EY027458 (YMP), Fight for Sight-International Retinal Research Foundation FFSGIA16002 (YMP), unrestricted departmental support from Research to Prevent Blindness, and the University of Michigan Department of Ophthalmology and Visual Sciences, and China Scholarship Council No.201806370270.

Conflict of interest

None.

Author details

Yanxiu Li[1,2] and Yannis Mantas Paulus[1,3*]

1 Department of Ophthalmology and Visual Sciences, University of Michigan, Ann Arbor, MI, United States

2 Department of Ophthalmology, Xiangya Hospital, Central South University, Changsha, P.R. China

3 Department of Biomedical Engineering, University of Michigan, Ann Arbor, MI, United States

*Address all correspondence to: ypaulus@med.umich.edu

References

[1] Zhang HF, Maslov K, Stoica G, et al. Functional photoacoustic microscopy for high-resolution and noninvasive *in vivo* imaging. Nature Biotechnology. 2006;**24**:848-851

[2] Laufer J, Zhang E, Raivich G, et al. Three-dimensional noninvasive imaging of the vasculature in the mouse brain using a high resolution photoacoustic scanner. Applied Optics. 2009;**48**:D299-D306

[3] de la Zerda A, Zavaleta C, Keren S, et al. Carbon nanotubes as photoacoustic molecular imaging agents in living mice. Nature Nanotechnology. 2008;**3**:557-562

[4] de la Zerda A, Paulus YM, Teed R, et al. Photoacoustic ocular imaging. Optics Letters. 2010;**35**:270-272

[5] Li C, Wang LV. Photoacoustic tomography and sensing in biomedicine. Physics in Medicine and Biology. 2009;**54**:R59-R97

[6] Wang LV. Hu S. photoacoustic tomography: *In vivo* imaging from organelles to organs. Science. 2012;**335**:1458-1462

[7] Zhang HF, Puliafito CA, Jiao S. Photoacoustic ophthalmoscopy for *in vivo* retinal imaging: Current status and prospects. Ophthalmic Surgery, Lasers & Imaging. 2011;**42**(Suppl):S106-S115

[8] Xing WX, Wang LD, Maslov K, et al. Integrated optical- and acoustic-resolution photoacoustic microscopy based on an optical fiber bundle. Optics Letters. 2013;**38**:52-54

[9] Nguyen V, Paulus YM. Photoacoustic ophthalmoscopy: Principle, application, and future directions. Journal of Imaging. 2018;**4**:149

[10] Liu W, Zhang HF. Photoacoustic imaging of the eye: A mini review. Photoacoustics. 2016;**4**:112-123

[11] Zhang W, Li YX, Nguyen VP, et al. High-resolution, *in vivo* multimodal photoacoustic microscopy, optical coherence tomography, and fluorescence microscopy imaging of rabbit retinal neovascularization. Light, Science & Applications. 2018;**7**:103

[12] Jiao S, Jiang M, Hu J, et al. Photoacoustic ophthalmoscopy for *in vivo* retinal imaging. Optics Express. 2010;**18**:3967-3972

[13] Hu S, Rao B, Maslov K, et al. Label-free photoacoustic ophthalmic angiography. Optics Letters. 2010;**35**:1-3

[14] Zhao H, Wang G, Lin R, et al. Three-dimensional hessian matrix-based quantitative vascular imaging of rat iris with optical-resolution photoacoustic microscopy *in vivo*. Journal of Biomedical Optics. 2018;**23**:1-11

[15] Rao B, Li L, Maslov K, et al. Hybrid-scanning optical-resolution photoacoustic microscopy for *in vivo* vasculature imaging. Optics Letters. 2010;**35**:1521-1523

[16] Liu W, Schultz KM, Zhang K, et al. *In vivo* corneal neovascularization imaging by optical-resolution photoacoustic microscopy. Photoacoustics. 2014;**2**:81-86

[17] Jeon S, Song HB, Kim J, et al. *In vivo* photoacoustic imaging of anterior ocular vasculature: A random sample consensus approach. Scientific Reports. 2017;**7**:4318

[18] Kelly-Goss MR, Ning B, Bruce AC, et al. Dynamic, heterogeneous endothelial Tie2 expression and capillary blood flow during

microvascular remodeling. Scientific Reports-UK. 2017;7:9049

[19] Tian C, Zhang W, Mordovanakis A, et al. Noninvasive chorioretinal imaging in living rabbits using integrated photoacoustic microscopy and optical coherence tomography. Optics Express. 2017;**25**:15947-15955

[20] Song W, Wei Q, Liu T, et al. Integrating photoacoustic ophthalmoscopy with scanning laser ophthalmoscopy, optical coherence tomography, and fluorescein angiography for a multimodal retinal imaging platform. Journal of Biomedical Optics. 2012;**17**:061206

[21] Song W, Wei Q, Feng L, et al. Multimodal photoacoustic ophthalmoscopy in mouse. Journal of Biophotonics. 2013;**6**:505-512

[22] Liu T, Li H, Song W, et al. Fundus camera guided photoacoustic ophthalmoscopy. Current Eye Research. 2013;**38**:1229-1234

[23] Hughes A. A schematic eye for the rabbit. Vision Research. 1972;**12**:123-138

[24] Delori FC, Webb RH, Sliney DH, et al. Maximum permissible exposures for ocular safety (ANSI 2000), with emphasis on ophthalmic devices. Journal of the Optical Society of America. A, Optics, Image Science, and Vision. 2007;**24**:1250-1265

[25] Wei Q, Liu T, Jiao SL, et al. Image chorioretinal vasculature in albino rats using photoacoustic ophthalmoscopy. Journal of Modern Optics. 2011;**58**:1997-2001

[26] Srinivasan VJ, Ko TH, Wojtkowski M, et al. Noninvasive volumetric imaging and morphometry of the rodent retina with high-speed, ultrahigh-resolution optical coherence tomography. Investigative Ophthalmology & Visual Science. 2006;**47**:5522-5528

[27] Song W, Wei Q, Jiao S, et al. Integrated photoacoustic ophthalmoscopy and spectral-domain optical coherence tomography. Journal of Visualized Experiments. 2013;**71**:4390

[28] Nguyen V, Li YX, Zhang W, et al. Multi-wavelength, en-face photoacoustic microscopy and optical coherence tomography imaging for early and selective detection of laser induced retinal vein occlusion. Biomedical Optics Express. 2018;**9**:5915-5938

[29] Ambati J, Ambati BK, Yoo SH, et al. Age-related macular degeneration: Etiology, pathogenesis, and therapeutic strategies. Survey of Ophthalmology. 2003;**48**:257-293

[30] Dai C, Li L, Liu W, et al., editors. In vivo time-serial evaluation of laser-induced choroidal neovascularization in rats simultaneously using photoacoustic microscopy and optical coherence tomography. Photons Plus Ultrasound: Imaging and Sensing. SPIE. 2018

[31] Khoobehi B, Firn K, Thompson H, et al. Retinal arterial and venous oxygen saturation is altered in diabetic patients. Investigative Ophthalmology & Visual Science. 2013;**54**:7103-7106

[32] Hardarson SH, Stefansson E. Retinal oxygen saturation is altered in diabetic retinopathy. The British Journal of Ophthalmology. 2012;**96**:560-563

[33] Olafsdottir OB, Hardarson SH, Gottfredsdottir MS, et al. Retinal oximetry in primary open-angle glaucoma. Investigative Ophthalmology & Visual Science. 2011;**52**:6409-6413

[34] Vandewalle E, Pinto LA, Olafsdottir OB, et al. Oximetry in glaucoma: Correlation of metabolic change with

structural and functional damage. Acta Ophthalmologica. 2014;**92**:105-110

[35] Song W, Wei Q, Liu WZ, et al. A combined method to quantify the retinal metabolic rate of oxygen using photoacoustic ophthalmoscopy and optical coherence tomography. Scientific Reports—UK. 2014;**4**:6525

[36] Liu WZ, Zhang HF. Noninvasive *in vivo* imaging of oxygen metabolic rate in the retina. In: Conference Proceedings: Annual International Conference of the IEEE Engineering in Medicine and Biology Society. IEEE Engineering in Medicine and Biology Society. Annual Conference. 2014. pp. 3865-3868

[37] Hariri A, Wang J, Kim Y, et al., editors. In Vivo Photoacoustic Imaging of Chorioretinal Oxygen Gradients. Journal of Biomedical Optics. SPIE; 2018

[38] Weiter JJ, Delori FC, Wing GL, et al. Retinal pigment epithelial lipofuscin and melanin and choroidal melanin in human eyes. Investigative Ophthalmology & Visual Science. 1986;**27**:145-152

[39] Schmidt SY, Peisch RD. Melanin concentration in normal human retinal pigment epithelium. Regional variation and age-related reduction. Investigative Ophthalmology & Visual Science. 1986;**27**:1063-1067

[40] Kanis MJ, Berendschot TT, van Norren D. Influence of macular pigment and melanin on incident early AMD in a white population. Graefe's Archive for Clinical and Experimental Ophthalmology. 2007;**245**:767-773

[41] Berendschot TT, Willemse-Assink JJ, Bastiaanse M, et al. Macular pigment and melanin in age-related maculopathy in a general population. Investigative Ophthalmology & Visual Science. 2002;**43**:1928-1932

[42] Lapierre-Landry M, Carroll J, Skala MC. Imaging retinal melanin: A review of current technologies. Journal of Biological Engineering. 2018;**12**:29

[43] Silverman RH, Kong F, Chen YC, et al. High-resolution photoacoustic imaging of ocular tissues. Ultrasound in Medicine & Biology. 2010;**36**:733-742

[44] Jiang M, Zhang X, Puliafito CA, et al. Adaptive optics photoacoustic microscopy. Optics Express. 2010;**18**:21770-21776

[45] Shu X, Li H, Dong BQ, et al. Quantifying melanin concentration in retinal pigment epithelium using broadband photoacoustic microscopy. Biomedical Optics Express. 2017;**8**:2851-2865

[46] Zhang X, Zhang HF, Puliafito CA, et al. Simultaneous *in vivo* imaging of melanin and lipofuscin in the retina with photoacoustic ophthalmoscopy and autofluorescence imaging. Journal of Biomedical Optics. 2011;**16**:080504

[47] Liu T, Wei Q, Song W, et al. Near-infrared light photoacoustic ophthalmoscopy. Biomedical Optics Express. 2012;**3**:792-799

[48] Liu X, Liu T, Wen R, et al. Optical coherence photoacoustic microscopy for *in vivo* multimodal retinal imaging. Optics Letters. 2015;**40**:1370-1373

[49] Shu X, Liu W, Zhang HF. Monte Carlo investigation on quantifying the retinal pigment epithelium melanin concentration by photoacoustic ophthalmoscopy. Journal of Biomedical Optics. 2015;**20**:106005

[50] Tian C, Zhang W, Nguyen VP, et al. Novel photoacoustic microscopy and optical coherence tomography dual-modality chorioretinal imaging in living rabbit eyes. Journal of Visualized Experiments. 2018;**8**:132

[51] Nguyen V, Li YX, Aaberg M, et al. *In vivo* 3D imaging of retinal neovascularization using multimodal photoacoustic microscopy and optical coherence tomography imaging. Journal of Imaging. 2018;**4**:150

[52] Wang T, Nandy S, Salehi HS, et al. A low-cost photoacoustic microscopy system with a laser diode excitation. Biomedical Optics Express. 2014;**5**:3053-3058

[53] Wang Y, Li CH, Wang RK. Noncontact photoacoustic imaging achieved by using a low-coherence interferometer as the acoustic detector. Optics Letters. 2011;**36**:3975-3977

[54] Rousseau G, Blouin A, Monchalin JP. Non-contact photoacoustic tomography and ultrasonography for tissue imaging. Biomedical Optics Express. 2012;**3**:16-25

[55] Chen J, Lin R, Wang H, et al. Blind-deconvolution optical-resolution photoacoustic microscopy *in vivo*. Optics Express. 2013;**21**:7316-7327

[56] Luke GP, Yeager D, Emelianov SY. Biomedical applications of photoacoustic imaging with exogenous contrast agents. Annals of Biomedical Engineering. 2012;**40**:422-437

[57] Yang XM, Stein EW, Ashkenazi S, et al. Nanoparticles for photoacoustic imaging. Wiley interdisciplinary reviews. Nanomedicine and Nanobiotechnology. 2009;**1**:360-368

[58] Zanganeh S, Li H, Kumavor PD, et al. Photoacoustic imaging enhanced by indocyanine green-conjugated single-wall carbon nanotubes. Journal of Biomedical Optics. 2013;**18**:096006

[59] Li W, Chen X. Gold nanoparticles for photoacoustic imaging. Nanomedicine (London, England). 2015;**10**:299-320

[60] Mallidi S, Larson T, Tam J, et al. Multiwavelength photoacoustic imaging and plasmon resonance coupling of gold nanoparticles for selective detection of cancer. Nano Letters. 2009;**9**:2825-2831